Phloem translocation

Phloem translocation

Phloem translocation

M. J. CANNY

Foundation Professor of Botany
Monash University

CAMBRIDGE
at the University Press 1973

Published by the Syndics of the Cambridge University Press
Bentley House, 200 Euston Road, London NW1 2DB
American Branch: 32 East 57th Street, New York, N.Y. 10022

© Cambridge University Press 1973

Library of Congress Catalogue Card Number: 72–83583

ISBN: 0 521 20047 4

Printed in Holland by B.V. D. Reidel Book Manufacturers, Dordrecht

Contents

Preface

This book is less a review of a field of plant physiology than a lengthy statement of a point of view about that field which, in the course of the argument, involves the review of most of the original work. It first attempts to set out those facts which are most surely known; comparing, checking and reducing to common terms the diverse observations of all whose work bears on the problem. Some errors are corrected; some observations are minimised as inconsistent with the main body of work; some neglected experiments are polished up for display. Throughout the first part a conscious effort has been made to keep the author's personal bias in check, and to let the logic of the facts present its own argument. This has not proved wholly attainable, and at times when this flow of facts has thrown up a correspondence that speaks loudly for my views I have not been able to resist patting it on the head. In the second part the single point of view takes over almost completely, and an attempt has been made to persuade the reader that nearly the entire range of experimental observations can be accommodated by a hypothesis; to show how much published work can be interpreted in terms of the hypothesis and in ways alien to the authors' intentions to provide confirmation of my view; how this hypothesis makes predictions; how some of these predictions are substantially verified, and how some remain to be tested. A few recalcitrant facts that do not fit the hypothesis are taken out, dusted and shaken, and put back where they may be examined again. In the course of the exposition a good deal of new experimental work is published for the first time: most of it my own done under the stimulus of the developing logic and aimed at supplying unknown data whose want became apparent; some of it that of colleagues, collected incidentally to their more serious concerns or under my prompting, and, being of limited value to its owner, now generously given to suffer my interpretation. The bibliography is not exhaustive, though I have tried to give credit for all the main original discoveries and thoughts, and have not consciously concealed any work because it was inconvenient. Deliberate exclusion was made of two large areas that seemed irrelevant: auxin transport, and the vast literature on movement of weedkillers. In so far as it has not been written for my own enjoyment, it is directed to botanists who want to become acquainted with the field, to begin work in some part of it, to use its techniques in their own work, or its ideas as a complement to their own.

M. J. CANNY

Monash University
September 1971

Acknowledgements

I gratefully record my thanks to those who have contributed largely to this book, but who are not acknowledged in the text.

First, there are those who by their interest and encouragement over the years, their questioning of ideas and facts, their kindly criticism and helpful suggestions, have supported this endeavour and removed many of its irregularities: Professor G. E. Briggs, Professor Denis Carr, Professor T. C. Chambers, Dr Stewart Cox, Professor Jack Dainty, Dr G. C. Evans, Professor David Fensom, Professor Adrian Horridge, the late Dr John Lawton, Mr Peter Lumley, Professor Robert Milburn, Dr Terry O'Brien, Professor Michael Pitman, Sir Rutherford Robertson, Professor Owen Phillips, Dr Stella Thrower (formerly Ovenden), Professor John S. Turner, Dr Catherine Whittle, Professor Harold Woolhouse, Mr D. R. Zeidler and Dr Martin Zimmermann.

Second, there are many whom I have consulted on specific points and who have given generously of their time and expertise. Principal among them were: Professor George Batchelor, Mr Bob Dewar, Dr John Edwards, Dr J. S. Kennedy, Dr Tom Neales, Dr Kingsley Rowan, Sir Geoffrey Taylor, Dr Robert Thaine, Dr Barry Tomlinson.

Third, there are those who have shared with me the daily labour of laboratory and desk and who have contributed far more than the work of their hands: Mrs Katalin Markus, Mr Michael Askham, Miss Margo Harvey (now Mrs Dodson), and Mrs Edith Hedley who has patiently typed so many drafts.

Finally there are those who have given financial assistance and general research facilities: The Universities of Cambridge and Monash, the research laboratory of Imperial Chemical Industries of Australia and New Zealand, The Royal Society and Nuffield Foundation who enabled me to visit Central Africa and see the sausage trees, and the Australian Research Grants Committee, who for the past six years have generously provided equipment and an assistant.

PART ONE

The experimental facts

1. The transfer of dry weight

One cannot fix one's eyes on the commonest natural production without finding food for a rambling fancy.

Mansfield Park

Consider a pumpkin. It weighs, let us say, 4500 g. Of this weight, part is water; how much, we can easily find by weighing a sample before and after drying. Suppose this indicates that the tissue is 90 per cent water, then the dry weight of the pumpkin is 450 g. Of this dry weight most is organic and combustible, a little is mineral and will be left as ash on burning; all of it has either been synthesised by the pumpkin or imported through the stalk. The mineral part must all have been imported, some of the organic part can be made by the photosynthetic activity of the growing pumpkin, and if we seek to know the total amount of dry matter that has entered the pumpkin we must subtract from the measured dry weight the photosynthetic products of the cells of the pumpkin, but add an amount of carbohydrate to allow for what has been respired to carbon dioxide by all the cells during growth. These two corrections, one positive and one negative, are probably small compared with the main movement of substance down the stalk, and moreover tend to balance one another. For example, Clements (1940) found that the growth in dry weight of the fruit of the sausage tree was not detectably reduced by enclosing the fruit cluster in a black bag, and estimated the respiratory loss as 10 percent of the transported dry weight. The dry matter of the fruit has arrived through the stalk; indeed the amount in the fruit is probably an under-estimate of what has been transported. It is this phenomenon of transport of organic substance by plants that is called translocation and which it is the purpose of this book to explore.

The phenomenon is seen most strikingly in the growth of fruits but it is a phenomenon that is as fundamental to the life of plants as is the circulation of blood to the life of animals. Everywhere in a plant that organic substance is being consumed, elaborated into cells, tissues and organs, stored away as food reserves; at apices, at meristems and cambia, substance must be arriving through the translocation system. These places are called *sinks*. Similarly, everywhere that substance is being produced in excess of the immediate requirements of the surrounding cells, in photosynthesising leaves, in sprouting tubers and germinating seeds, in stems from which reserves are being mobilised, substance is being carried away by the translocation system. These places are called *sources*. The translocation system connects sources to sinks and carries organic substances from the one to the other. Growing roots and shoots can be regarded as self-producing tubes that carry their substance to the meristem at the end where it is made into tube wall, extending the tube and carrying the building unit forward. The core of these tubes is the translocation system bearing the substance of the tube in solution. The two-way traffic of the system can be appreciated in leaves which start as sinks and are formed from substance imported by the translocation system, but when fully-grown become sources, exporting substance to other sinks through the same system.

Science works by making comparisons and the first step in any scientific study is to measure the phenomenon – to ascribe units and to assign numbers of these units to various examples of the pheonomenon in order that comparisons may be made. Our pumpkin took, we may suppose, about six weeks to attain its final size, giving a rate of movement of dry matter through the stalk of 450 g per $6 \times 7 \times 24$ hours. If it were certain that all the tissues of the stalk carried on the transport, we might divide our figures by the cross-sectional area of the stalk and ascribe our units as g dry wt hr^{-1} cm^{-2} stalk. But it is certain that some cells of the stalk are not involved in the transport, and probable that only a few are, making the whole area irrelevant for our purpose of comparison. The next chapter will be devoted to showing that the path of translocation is in the phloem and to discussing the other views that have been held and the reasons why they must be rejected. For the present we will take it on trust that the path is in the phloem. 'In' the phloem, you will notice, for it will be seen in Chapter 2 that the path may involve not the whole phloem but only the specialised sieve elements. We may write our unit of translocation as g dry wt hr^{-1} cm^{-2} phloem and will need to know the area of phloem (sieve elements, companion cells and associated parenchyma) in the cross section of the pumpkin stalk. Microscopic measurement gave this as 0.146 cm^2 giving the rate of translocation as 3.06 g dry wt hr^{-1} cm^{-2} phloem. Most workers have taken the sieve elements to occupy one fifth of the total area of the phloem, so if the path proves to be limited to the sieve elements only we should have to multiply our figures by about five and write our units as g dry wt hr^{-1} cm^{-2} sieve element. This fraction of one fifth will require further consideration, and for the moment the whole phloem is taken as the path; but let the reader note that this is the only assumption that has so far been made. All other variables are unequivocally measurable.

The number of published measurements of this kind, assessing the rate of transfer of dry weight into fruits, is not so great that we cannot consider them all, but before doing so it will be necessary to explain the further assumption that most authors have introduced in giving the results of their experiments. They have assumed: (1) that the transfer takes place as sugar in solution; (2) that the solution has a certain concentration of dissolved sugar; and (3) that the solution is moving at a certain speed. Then, these quantities will be related by an equation:

$$\text{mass transfer} = \text{concentration} \times \text{area} \times \text{speed}$$
$$\text{g hr}^{-1} \qquad \text{g cm}^{-3} \qquad \text{cm}^2 \qquad \text{cm hr}^{-1} \qquad (1.1)$$

This is not a law or a statement of fact, but a definition of a hypothetical solution, which, if it has the properties shown on the right-hand side, will result in the mass transfer shown on the left-hand side. The mass transfer is the only certain measurement; the other quantities are all guessed. More usually, two are guessed, concentration and area, and the speed is derived, but it remains hypothetical. The identification of the moving organic substance with sugar is another assumption. This has excellent experimental warranty which will be taken up in detail in Chapter 2. For the present it may be regarded as a fact. The previous unit contained the single assumption that the area of the path is known; we may give the quantity a name and relate it to the properties of the hypothetical solution by the equation:

$$\text{specific mass transfer} = \text{concentration} \times \text{speed}$$
$$\text{g hr}^{-1} \text{ cm}^{-2} \qquad\qquad \text{g cm}^{-3} \qquad \text{cm hr}^{-1}$$

The fashion for expressing results of translocation measurements as speeds was set by Dixon and Ball (1922) when recording the measurement of rate of transfer of dry weight into a single growing potato tuber. The dry mass was 50 g, the time for its growth was 100 days, and the phloem area was 0.00422 cm². These figures give a specific mass transfer of 4.5 g hr^{-1} cm^{-2} phloem but were expressed by Dixon and Ball (1922) as being achieved by flow of a 10 per cent solution of sugar at 40 cm hr^{-1}, and used in this form as an argument that the phloem could not possibly be carrying on transport at this rate and that other tissues must be involved. Measurements of this kind thus became involved in the general controversy over the channel of movement and the fashion of expressing results as 'velocities' persists until this time.

A similar set of figures is published by Mason and Lewin (1926) on the growth of tubers of the yam, *Dioscorea alata*. In this experiment, as in that of Dixon and Ball, the photosynthetic contribution of the fruit is not present since the growing organ was underground. Their figures (Table 1.1) lead to a specific mass transfer of 4.4 g hr^{-1} cm^{-2} phloem, but were expressed by them as requiring the flow of a 25 per cent sugar solution at 88 cm hr^{-1} through the sieve elements (regarded as one fifth

TABLE 1.1 *Values of the rate of specific mass transfer of dry weight derived from published data*

Author	Plant	Number	Dry weight mass g	Time days	Phloem area cm²	Specific mass transfer g hr⁻¹ cm⁻² phloem
Stems, tuber and fruit stalks						
Dixon & Ball (1922)	*Solanum tuberosum* tuber	1	50	100	0.00422	4.5
Mason & Lewin (1926)	*Dioscorea alata* tuber	43	42.7 (max.) 11 (av.)	7	0.057	4.4 1.15
Crafts (1933)	*Solanum tuberosum* tuber	140	18.7 (max.)	21	0.0176	2.1
Clements (1940)	*Kigelia africana* fruit	1	32.5 (max.)	1	0.53	2.6
Crafts & Lorenz (1944)	Connecticut field pumpkin	39	482	33	0.186	3.3
	Early prolific straightneck summer squash	36				0.85 to 3.77 mean 2.68
Colwell (1942)	*Cucurbita*	5	807	28	0.250	4.8
Evans *et al.* (1970)	Wheat peduncle	av.	Regression line of mass transfer on phloem area, Fig. 1.1			4.4
Tammes (1933, 1952)	*Arenga* cut flower stalk	av.	33.6	1/24	0.34	99
Münch (1930)	*Pirus communis* fruit	1	5.93	120	0.0035	0.59
	Prunus domestica fruit	1	3.85	92	0.00308	0.57

Table 1.1 (Continued)

Author	Plant	Number	Dry weight mass g	Time days	Phloem area cm²	Specific mass transfer g hr⁻¹ cm⁻² phloem
	Fagus sylvatica fruit	I	2.22	120	0.00127	0.61
	Quercus pedunculata fruit	I	1.63	45	0.00254	0.59
	Quercus peduncula-ta × sessiliflora fruit	I	3.04	45	0.00594	0.47
Tree trunks						
Münch (1930)	*Pinus sylvestris* I	I	3020	120	0.984	1.07
	Pinus sylvestris II	I	1270	120	0.374	1.18
	Acer pseudoplatanus	I	4790	120	0.263	6.33
	Acer platanoides	I	7990	120	0.538	5.15
	Tilia parvifolia	I	3830	120	0.455	2.92
	Carpinus betulus	I	3890	120	0.216	6.25
	Quercus rubra	I	8730	120	0.674	4.50
Crafts (1931a)	Pear tree	16	6.53	I	0.30	0.9
Petioles						
Birch-Hirschfeld (1920)	*Phaseolus multi-florus*	av.	0.003276	1/24	0.0046	0.56
Dixon (1932)	*Tropaeolum majus*	av.	0.31	I	0.0009	1.4
Crafts (1931a)	*Tropaeolum majus*	5	0.237*	I	0.01945	0.51
	Phaseolus	40	0.0061	10/24	0.00123	0.5
Sachs (1884)	*Helianthus annuus*	av.	0.97	1/24	0.21	4.6
Geiger et al. (1969)	*Beta vulgaris*	13	0.0014	1/24	0.00102	1.37 ± 0.13

* An error in Crafts gives 0.327.

of the phloem). This was the maximum rate of transfer, achieved from the 31st to the 35th week of growth, the steepest slope of a sigmoid curve produced by plotting dry weights of tuber against time. A little consideration would have warned us to expect such a curve of fruit weight against time; it is most likely that the rate of specific mass transfer will vary during the growth of the fruit from low values at the beginning and end to a high peak in the middle of growth. Most authors have been concerned to measure the maximum rate of transfer, not only because it is the most spectacular, but also because it is the most difficult to explain and any theory of translocation must be capable of accommodating the highest measured values. The maximum rate has not always been distinguished from the average rate taken over the whole growth period as in Dixon and Ball's experiment. The average rate of specific mass transfer into the yams of Mason and Lewin was $1.15 \text{ g hr}^{-1} \text{ cm}^{-2}$ phloem. A further complication, and one which detracts from the usefulness of a long-period average value of specific mass transfer, arises in the possible growth of the phloem of the stalk along with the growth of the fruit. Then the area of the path by which the dry-weight transfer is divided will be varying. Mason and Lewin specifically state that the phloem of the yam stem was fully grown before the yam started to grow, but in other fruits the change of phloem area may be considerable. Taking all these factors into consideration, some short-period average of specific mass transfer seems the most satisfactory measure of translocation, and we proceed to compare the remaining published values on this basis.

Münch (1930) measured the dry weights of some tree fruits, pear, plum, beech and acorn, and assessed their growth as transport through the phloem of the fruit stalk. He apparently measured only one fruit of each, but the values are closely congruent. They yield time average transfer rates about half those into the underground storage organs.

Crafts (1933) provides more complete data than do Dixon and Ball on the rate of transport into potato tubers, and expresses his results on a new basis, that of rate of movement of solid sugar (using a value for the density of sugar of 1.5). The rate calculated from an average of 140 tubers is for a week's translocation at the time of rapid import, and the figures reproduced in Table 1.1 lead to a value of specific mass transfer of $2.1 \text{ g hr}^{-1} \text{ cm}^{-2}$ phloem.

Clements (1940) supplies data for a more unusual fruit, the spectacular hanging gourds of the sausage tree, whose attractions can be gauged as well from a laconic sentence as from Fig. 16.2: 'Two of these (clusters) were interfered with by tourists and were abandoned.' His results exemplify several of the points already discussed: the sigmoid time-curve of dry-weight increase, the changing activity of the floral stem, and the effects of fruit photosynthesis and respiration; but are new in that an attempt is made to measure the concentration of the moving solution. If this could be done it would remove one of the pieces of guesswork from the mass transfer equation (1.1) but Clements's results show how difficult the measurement is. Phloem tissue was dissected from the floral stem and analysed for sugars. The total sugar content of the phloem was expressed as

$$\frac{total\ sugar}{total\ water\ content} \times 100$$

giving values of 2 to 3 per cent. Clements then proceeds to argue that the rate of flow of such a weak solution (or in fact of the still weaker solution of reducing sugars which for some reason he considers the important sugars for transport) would have to move at impossible rates to achieve the measured mass transfer. The conclusion we may draw from the measurements, however, is that the method did not give a close estimate of the concentration of the moving solution. There must be much water in the phloem tissue that is not moving with the sugar. Nevertheless, for all the inadequacy of the concluding argument, the attempt to estimate the solution concentration represented a clear advance in thought about the problem.

Two estimates of rates of dry-weight transfer into *Cucurbita* fruits are to be found in a paper by Crafts and Lorenz (1944), their own and some measurements of Colwell contained in a thesis and unpublished save for this citation. The results are given in the form suited to the chosen unit in Table 1.1. They are closely similar to the others discussed and present no new concepts. To avoid the photosynthetic complication a non-chlorophyllous variety (Early Straightneck) of squash was used in one set of measurements.

The constancy of the specific mass transfer value suggests that there is probably a close correspondence between the capacity of an organ to translocate and the amount of phloem it contains, or, to put it the other way round, that the rate of translocation may be limited by the area of phloem available. Evans, Dunstone, Rawson and Williams (1970) tested this idea by measuring the rate of translocation into ears of wheat in a series of lines and varieties that had a great range of areas of phloem in their peduncles. Some of this variation was achieved by using different species of diploid, tetraploid and hexaploid varieties of *Triticum* and *Aegilops*; some by manipulating the hexaploid species *Triticum aestivum*, which responded to longer periods of vernalisation by producing fewer spikelets and less phloem in the peduncle. The (maximum) translocation to the ears was assessed by measuring the increase in dry weight, subtracting a correction for assimilation by the ear (determined by gas analysis in a chamber), and adding a correction for dark respiration. The results are shown in Fig. 1.1. The regression line fitted to the points (continuous line) gives a value of the specific mass transfer of 4.4 g hr^{-1} cm^{-2} phloem, and the correlation coefficient is 0.86. A cursory reading of the original paper would suggest that their measurements led to an estimate of 3.3 g hr^{-1} cm^{-2} phloem, the slope of the line on their graph (interrupted line of Fig. 1.1), but closer examination shows that the latter is derived solely by substituting into our equation 1.1 the concentration

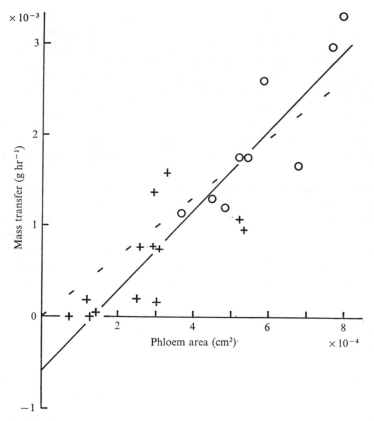

Fig. 1.1. Maximum mass transfer to wheat ears plotted against phloem area of the culm. The large range of phloem areas was obtained by: (*a*) choosing diploid and tetraploid wild wheats (+); and (*b*) manipulating the ear-size of the hexaploid *Triticum aestivum* by different periods of vernalisation (○). Data of Evans *et al.* (1970) Fig. 2, re-drawn. The continuous line is the regression of mass transfer on phloem area. The interrupted line is the line from the original figure which is calculated from particular values of the parameters of equation 1.1.

10 per cent, the speed 100 cm hr^{-1}, and assuming that the sieve tubes, carrying all the flow, occupy one third of the phloem. The suggestion that translocation is limited by the quantity of phloem available, and that the tissue has a uniform transport capacity, receives strong support from this consistent set of data.

From all these measurements summarised in Table 1.1 we may conclude that translocation into fruits can reach a maximum rate of about 4 g dry wt hr^{-1} cm^{-2} phloem, and for much of the time of growth of a fruit proceeds at 1 to 2 of these units. The range of values appears to be quite small. We must be cautious, however, in drawing any inference from such measurements about a possible speed of translocation, for the value derived for the speed will depend upon the assumptions made.

Formally similar, but concerning transport to a different kind of sink, is the calculation of Crafts (1931*a*) for movement in the bark of two-year-old pear trees. The sink here was the whole trunk and roots below the crown of leaves, and the amount of matter arriving was assessed by measuring the increasing dry weight of 16 trees 'uniform in size and shape', sampling two at a time over a period from 1 August to 18 December. The phloem area in the bark at the top of each trunk was measured at harvest and contributed to an average of 0.30 cm^{2} per tree for the whole set. The daily increase in dry weight per tree per day averaged 6.53 g. The specific mass transfer is therefore $6.53/(24 \times 0.30) = 0.9$ g hr^{-1} cm^{-2} phloem.

This kind of calculation had been done by Münch (1930) for larger trees where it is of course more complicated. He cut them into 2 m lengths and measured the width of the five outer rings and assessed from these the average annual volume growth of wood. The specific gravity of the material of the last annual ring was measured, and applied to the volume figures to get the dry weight transported down into each annual ring. An amount was added to this to allow for the weight of the roots and bark: 45 per cent of the stem wood, in accordance with foresters' practice. Then, with a measure of the area of active phloem at the base of the crown, and a growth period of 120 days, he calculated specific mass transfers, though in other units than ours. He goes on to turn these into speeds of flow in the contemporary fashion. His values agree well with all the rest.

The record for rate of movement is set by a damaged system, the bleeding flower stalk of the palm *Arenga* whose inflorescence had been cut off. This system occupied the attention of Tammes in the Minahassa during the early 1930s:

In the tropics a sugar-containing juice is tapped from various species of palms; large quantities of this juice being used locally for the manufacture of sugar and alcoholic beverages, the so-called toddy.

The tapping of this juice may take place in different ways. All methods, however, have in common that a bruise or a wound is made at the spot where the plant is to be tapped. Generally only the inflorescences of the trees are tapped, e.g. *Cocos*, *Nipa* and *Arenga*. In some cases, however, the stem is tapped, e.g. *Phoenix*.

The quantity of juice obtained is very variable. *Nipa* may yield about 1 liter of juice and *Cocos* about $1\frac{1}{2}$ liter in 24 hours. *Arenga* may attain 10 liters. Gibbs even mentions 42 liters per 24 hours for an inflorescence of *Corypha utan*. Usually, however, the inflorescences yield a much smaller quantity. (Tammes, 1933.)

The juice was a solution of 15–20 per cent sucrose with other minor components. It was not a xylem sap since the inflorescence still bled when removed from the palm. In a later paper (1952) Tammes records measurements of the phloem area of *Arenga* stalks, which enable us to calculate a specific mass transfer figure:

When, on an average, about 240 cc per hour flows out (5–6 liter per day) (Tammes 1933) and the diameter of the stalk is about 4,2 cm, the cross section of phloem elements will be 0,34 cm^2 and the rate of flow therefore about 7 meters per hour, or 11,5 cm per minute. The concentration of the sap is $\pm 14\%$ sucrose and the translocation of sap can therefore be calculated on 4,7 grams of sucrose per square cm of phloem per minute.

There is an error here in applying the area; the specific mass transfer is obviously $(240 \times 0.14)/0.34 = 99$ g sucrose hr^{-1} cm^{-2} phloem. The fact that this figure is thirty or so times larger than any other measurement of specific mass transfer should make us approach it with caution, particularly since in this system the sink is an artificial one where the natural sink is removed and the concentration of sugar is reduced to zero at the cut. The figure is recorded here as the exception emphasising the congruity of the other measurements.

So much for the rate of arrival of substance at sinks; we will now turn to a second analogous kind of measurement of translocation, that of the rate of movement of the products of assimilation out of leaves, movement away from a source. There are fewer of this kind of measurement and those we have are less satisfactory, having often been made for some other purpose. In principle the measurement is made by comparisons of the dry weights of attached and detached leaves during a period either in the light or in the dark. In the dark the change in dry weight of the attached leaves is the sum of respiration and translocation (both negative), and of the detached leaves, respiration only. Therefore

translocation = decrease in attached leaves – decrease in detached leaves.

In the light the changes are:

Detached	Attached
+ photosynthesis	+ photosynthesis
− respiration	− respiration
	− translocation

and

translocation = gain in dry wt of detached leaves − gain in attached leaves. No assumption is necessary that light and dark respiration are the same. The accuracy of the method depends on the extent to which the two implicit assumptions are justified: that the samples of leaves are initially strictly comparable; and that the only effect of detaching leaves is to prevent translocation. The first may be enforced by careful attention to the selection and number of samples; the second assumption is beyond control and the variable extent to which it may apply probably accounts for the variable results the method has yielded. Goodall (1946) discusses the available information on the second assumption and shows the extent of the differences in tomato plants.

Data collected around the turn of the century on the rate of assimilation of leaves had pointed the way to measurements of this kind. Differences in the accumulation of dry weight in the light between attached and detached leaves had left a substantial fraction of the dry-weight changes attributable to removal by translocation. Thus detached leaves in the light accumulated dry weight at a rate of about 1 g hr^{-1} m^{-2}* (Thoday, 1910b and those quoted by him, see below), while attached leaves increased in dry weight at a rate around 0 to 0.2 g hr^{-1} m^{-2}. Respiration accounted for about 0.05 of these units. All the authors dealing with assimilation had talked of the difference as the amount translocated, but Birch-Hirschfeld (1920) was the first to express the transport in a rational unit. Taking a reasonable average value for the net assimilation of *Phaseolus multiflorus* leaves as 0.5 g hr^{-1} m^{-2}m she argued that the leaves lost by translocation in 24 hours all the assimilate gained in 10 hours illumination, leaving no net change in weight over the day. The translocation rate per square metre of leaf surface was thus 0.28 g hr^{-1}. She measured the area of phloem in the petiole cross section as 0.0046 cm^2 for a leaf surface of 117 cm^2, so the rate of specific mass transfer was $0.003/0.0046 = 0.65$ g hr^{-1} cm^{-2} phloem. The value is substantially less than those derived for transfer into developing fruits and sets the standard for the other measurements of this class that are available. Thus, Dixon (1923) made a similar calculation:

* For the more customary modern unit of mg dm^{-2} hr^{-1}, multiply by 10.

Various investigators, from Sachs onwards, have measured the rate of photosynthesis per square meter of leaf per hour. Under the most favourable conditions the amount may approach 2 g, and it has been estimated as low as 0.5 g. Taking Brown and Morris' determination for *Tropaeolum majus*, viz., 1 g per square meter per hour, and assuming one third of the carbohydrate formed is used in respiration in the leaf, we find that a leaf of 46 cm^2 may form during ten hours' sunshine 0.46 g; during the twenty-four hours one third of this will be respired, leaving 0.31 g to be transported from the leaf. The volume of the solution (again assuming a concentration of 10 per cent) will be 3.10 cm^3. The cross-section of the bast of the bundles in the petiole was 0.0009 cm^2, therefore the velocity of flow, if the bast was used as the channel of transport, must have been 3.10/0.0009 × 24 or 140 cm per hour.

The attentive reader will have noticed that Dixon has taken 1000 cm^2 in a square metre, not 10,000, giving him an answer ten times what it should be. In our units, the specific mass transfer is 1.4 g hr^{-1} cm^{-2} phloem. His value for the phloem area is small compared with that found in the same plant in the next example, and the value of specific mass transfer correspondingly higher.

Crafts (1931a) repeats calculations like those of Dixon and Birch-Hirschfeld and measures his own *Tropaeolum majus* leaves to find a specific mass transfer of 0.51 g hr^{-1} cm^{-2} phloem. (Measured values in Table 1.1.) Not content to take a value for assimilation from other workers he went further and carried out the experiment outlined above with leaves of *Phaseolus*. This was probably the first measurement of this type made expressly for the purpose of finding a rate of translocation through a petiole and recording all the relevant data. Translocation during the 10 hours' light proceeded at a rate of 0.2 g hr^{-1} m^{-2}, and each square metre of lamina was correlated with 0.401 cm^2 of petiole phloem, giving a specific mass transfer of 0.5 g hr^{-1} cm^{-2} phloem. The rate in the dark (temperatures unspecified) was much less, about 0.11 of these units.

These results, agreeing closely with the calculation of Birch-Hirschfeld, might encourage us to believe that all petiole export was going on at a rate around 0.5 g hr^{-1} cm^{-2} phloem, but that calculations can be made on some of the old assimilation data for *Helianthus* leaves which show that in these large robust organs rates can be achieved which rival those quoted for import into fruits. For example, Sachs (1884) gives data for the changes in dry weight of *Helianthus* leaves in both light and dark. In the light at 24 °C attached leaves gained 0.68 g dry wt hr^{-1} m^{-2}, while the corresponding detached leaves gained 1.648 g hr^{-1} m^{-2}, giving a dry-weight loss of 0.97 g hr^{-1} m^{-2}. Sachs did not measure the phloem in the petioles but this can be readily done and an approximate figure used to calculate the phloem translocation. A graph of leaf area against petiolar phloem area in *Helianthus annuus* (Canny, unpublished) provides an average of $1.05 \pm 0.1 \times 10^{-2}$ cm^{-2} phloem per 500 cm^2 leaf surface, a figure from which the plants used by Sachs probably did not depart too widely. The specific mass transfer derived from the combined figures is $0.97/0.21 = 4.6$ g hr^{-1} cm^{-2} phloem, a figure as high as any on the fruit list. Other workers record rates of assimilation for *Helianthus* leaves of the same order: Brown and Morris (1893), about 1 g hr^{-1} m^{-2}; Brown and Escombe (1905), 0.5; Thoday (1910a & b), 1.65, who also gives reasons why the values found by Sachs, though perhaps high, are not inordinately so. Thoday reproduces a striking set of inaccessible data collected by Broocks for the daily march of dry weight of sugar beet leaves. Dry weight increases up till noon and then declines rapidly and uniformly until the following midnight or dawn. The sudden change at noon is ascribed to closure of the stomata stopping assimilation. Nomoto and Saeki (1969) resurrected this old technique, with the added refinement of steaming the petioles to stop translocation while the leaves were still attached, and presumably less disturbed than by cutting. They relied, perhaps unwisely, on short sampling periods (2 hours) to reduce the shrinkage errors that Thoday showed were serious in 5 hours. Their measured rates of assimilation of *Helianthus* are up to twice those of Sachs and comparable with those measured by gaseous exchanges in modern leaf-chamber work. Their data for translocation are incomplete, but show the major loss of dry weight occurs in the early afternoon, and reaches a peak value of about 1.3 g hr^{-1} m^{-2} in the period 12.00 to 14.00. If the same phloem measurement is applied to this as to Sach's measurements, a value of specific mass transfer of 6.2 g hr^{-1} cm^{-2} phloem is obtained. It seems probable that most small leaves export at a rate around 0.5 g hr^{-1} cm^{-2} phloem but that some exceptional leaves may do so at nearly ten times this rate for short periods.

An elegant variation of this method has been developed by Geiger and his co-workers (Geiger and Swanson, 1965a & b; Geiger, Saunders and Cataldo, 1969) which makes use of tracers. A young sugar beet plant is pruned down to a single source leaf and a single (younger) sink leaf connected through the stem/root complex. Carbon dioxide with a constant proportion of $^{14}CO_2$ is fed to the lamina of the source leaf and in about 100 minutes a steady-state labelling of the main pools accessible to the labelled carbon is achieved, as evidenced by a constant specific activity of the extracted sugars. Now the rate of arrival of this sugar at sinks along the path and at the end can be measured by dividing the rate of accumulation of label by the known specific activity. In the third of the cited papers the method is used to estimate a value of the specific mass transfer out of the beet petiole which is also included with the other petiole measurements of Table 1.1. The value of 1.37 ± 0.13 for an average of 13 different petioles fits well with the other estimates. In calculating this result a conversion factor of 12 grams of carbon per 27 grams of dry weight has been used (Thoday 1910*a*). Geiger *et al.* provide estimates of sieve-tube area as well as phloem area in the petioles, but the discussion of this complication is deferred until Chapter 11. They also make comparisons with speed and concentration in the usual way.

In addition to these two methods of measuring movement, arrival at sinks and departure from sources, there is a third way that has been used once; measurement of the changes in the channel when a blockage is made and translocated substance accumulates. The objection to such methods is that in making the blockage by some cut or interruption of the path we may have interfered with the rate of movement established before the cut was made; nevertheless, the measurements provide valuable additional data that are clearly comparable with the other kinds. The example of measurements of this kind is the data of Mason and Maskell (1928*a* & *b*) from which rates of transfer of carbohydrate in the stem of cotton were calculated. We shall have occasion to return to these important measurements several times, but at the moment may merely note the method and the range of values obtained. Measurements of sugar content of samples of the bark of cotton were made at intervals in the course of the day. The time interval was short enough for changes in soluble sugars to be reflecting translocation; the sugars seem not to have been built into the more complicated components of 'dry weight' which we have so far considered. The bark samples were delimited before measurement as strips and flaps by cutting the bark down to the wood and the channel was interrupted by a transverse cut so that it may be damaged to some unknown extent. For example the strip of bark below a leaf was separated by cuts from the surrounding bark and severed at the lower end. Then the products of assimilation that would in the intact plant have passed through this area of bark accumulated in it. Experimentally the measurements were achieved by sampling large numbers of similar plants, all with the same bark strip isolated. The change in sugar content of such a strip during a period gave a rate of transfer through the phloem area at the leaf-end of the strip. Many values of specific mass transfer are available from these experiments under different conditions, and are expressed in the later paper in the unit we have adopted (1928*b*, Table xxvi, col. 10). Notwithstanding the fact that they call these values 'Rate of Transport, grm per 1 cm^2 per hr in sieve-tubes', the area of the path is in fact what we have called 'phloem'. They state: 'The relative area of the following tissues in each zone (was) estimated: Cortex, primary and secondary ray tissues, fibres, and sieve-tube groups (including companion cells). For convenience we shall refer to the latter area as sieve-tubes even though companion cells were included.' The values range from 0.08 to 0.64 g hr^{-1} cm^{-2} phloem, and are closely correlated with the measured gradient of sugar concentration, a point which will be taken up in Chapter 3.

From the measurements discussed a clear picture emerges of the rate at which movement of organic substances takes place. In small herbaceous petioles and stems rates of specific mass transfer of 0.3 to 1 g hr^{-1} cm^{-2} phloem can be expected, while in large and vigorous organs and in the peduncles of large growing fruits the rate may be 1 to 4 of these units. We lack measurements of rates of movement in many classes of organ, notably in roots.

This transport process, on which the growth and elaboration of higher plants entirely depends, will be explored in the pages that follow. Most of the book will be devoted to establishing the rules that the process obeys, as this chapter has set a figure on the dimensions of the transport. These rules are in various degrees controversial, and the process of arriving at them will be by the comparison of what experimental data is available, testing the assembly of data for internal consistency and setting aside those measurements that seem unreliable either from their disagreement with the rest, or some internal methodological unsoundness. So, in the comparison of dry weight transfers, it seems reasonable to discard the palm inflorescence as having little directly in common with all the rest. It seems likely that some special circumstances are contributing here to a magnitude of movement that does not occur in all the others measured. As the data accumulate, and the rules are formulated, hypothetical mechanisms to explain what is happening will be treated. These hypotheses have had powerful directing influences on the measurements made, canalising for decades the work and thought into particular fashionable paths, and causing the neglect of other kinds of measurement and alternative viewpoints. Thus it has already been necessary to refer to the hypothesis of mass flow of sugar solution as a dominating fashion in the studies of dry-weight transfer. As the data and rules accumulate the other hypotheses to explain translocation will be studied and related to each other and to the experimental findings. Finally, the reasons will become plain why I personally favour a particular class of hypothesis, and believe that this at present is best in accord with the evidence; and that there are discrepancies between even this and the evidence which remain to be resolved, either by a new hypothesis or by revising the evidence.

The dry matter that moved into the pumpkin was very largely composed of carbohydrate, and the other movements considered were also carbohydrate movements. Other substances moved too, but in amounts which are so small as to relegate them to minor places in the consideration of the problem. It is with these major movements of carbohydrate that we have to deal, taking up first the points that were laid aside earlier, the twin problems of the channel in which the movement takes place, and the identification of the moving material as sucrose.

2. The channel of movement – including a discussion of the carbohydrates moving

Such experiments merely show the tremendous resistance of plants to mutilation but so far as depicting the process of whole plants is concerned, they show nothing.

Clements (1934)

There are two classical methods by which information has been sought on what is the channel in which the carbohydrates move: one is to interrupt the various possible channels and see how the movement is affected; the other, by analysis of the dissected parts of the plant to discover where changes in carbohydrates are occurring that are correlated with the movements. Experiments of the former kind are always open to the criticism that the operation of interruption, by cutting, steaming or poisons, has upset not only the tissue that was intended but nearby ones as well. The force of this objection has been all the more cogent because the two candidates for selection, the xylem and the phloem, are always so close together that it is difficult to interrupt one without interfering with the other. Experiments of the second kind are complicated by the fact that a consideration of the channel in which organic substances move cannot be separated from a knowledge of what chemical substances are moving. Movement is not to be measured in the abstract, it must be movement of something; the channel of movement is known to be the channel because those substances that are moving are found there rather than elsewhere and in amounts that vary as the movements vary. Similarly, in order to discover what is moving we must know where it is moving. Discussions of these matters are therefore necessarily mutually involved, and are in danger of becoming circular. The knot of this confusion has been cut in recent years by the application of a new technique, the tracing of labelled molecules, so that now it is easier to find either what is moving or where independently, though the distinction between substance moving and substance that has moved and accumulated is still very difficult.

Labelled molecules present in a particular tissue must have come there from the place they were applied; the first channel in which they spread must be the channel of movement. Again, labelled atoms appearing in a particular substance must be derived from the labelled substance applied; if one labelled substance predominates in the label that first spreads from the point of application, this is the substance that is moving. It is to the credit of plant physiology that the solution of this double problem, the channel of translocation and the form of the translocated substance, was practically completed by the exercise of labour and ingenuity ten years before the possibility of solution by the simpler method became known. Analysis with labelled molecules has confirmed and extended the conclusions already reached but, in this aspect of the subject, has provided little knowledge that is wholly new. The problem is conveniently treated historically, showing what was successively believed, and on what grounds.

The observations and opinions of generations about sap in plants were set down by the early writers: Grew, Malpighi, Perrault, Hales, Cotta, Knight, etc., that sap went up in the wood and came down in the bark. The role of the bark was apparently made clear by the reactions of a tree to removal of a ring of bark. The wound healed most rapidly above the ring, the upper lip swelled more than the lower, roots regenerated most easily from the upper side; sap was coming down from above. But if the stem below the ring was furnished with leaves like the stem above, there was less or no difference in the behaviour of the two sides of the wound; sap was spreading from the leaves through the bark. These observations led to a clear concept of physiological functioning: (1) the leaves wrought changes in the sap that were of the first importance for growth and the production of new tissues and organs; (2) the sap from the ground carrying some of the nourishment moved up the wood to the leafy part of the plant; (3) this sap, elaborated by the action of the leaves, moved to sites of growth and synthesis through the bark and any interruption of the bark prevented this flow and the subsequent changes and growth; (4) there were two separate streams of sap that travelled in distinct tissues and different directions. A good account of this early writing, with specific references, may be found in Hanstein (1860) and Esau (1961).

Hartig, who had first described sieve elements from phloem, published (1858) a body of experimental work confirming and extending the old ideas, using for his criterion of movement the accumulation and depletion of starch reserves in trees. Reserve materials were depleted from the side of the ring away from the source (below) and accumulated above the ring. The changes he observed were slow, often requiring a year for their manifestation, but the reliability of the classical picture was established by these and other contemporary researches. Hartig also recorded (1860) a confirming fact whose profound significance he plainly stated: that on making incisions in summer and autumn in the bark of many trees a few drops could be obtained of a solution of 25–33 per cent sugar (together with some minerals and nitrogen), which was different from the other two known plant saps, the xylem sap and the milky latex. He showed that the solution came from the part of the bark rich in sieve elements and suggested that it was probably nutrient sap descending from the leaves. Hanstein (1860) further says that it is the continuity of the sieve tubes in the phloem that is essential to the movement of the elaborated sap, and that these must be the channel. If pressed for details of the mechanism of the movement, the physiologists of the day would probably have followed Sachs in invoking diffusion down a gradient of concentration, a mechanism whose inadequacy is plainly stated by de Vries (1885) who suggested as an alternative that streaming protoplasm was doing the work. Sachs (1863) proposed that the pathway for carbohydrate was the starch sheath (leaving the phloem to carry nitrogenous materials), on the evidence that a high concentration of starch was found there. This is as good an example as could be found of the circularity of arguments about path and substance we have already noted, and the hypothesis did not survive the criticism of Heine (1885) and others. Czapek (1897) tried to confirm the phloem as the channel by showing: (1) that operations interrupting the direct phloem path but leaving oblique bridges of ground parenchyma stopped the movement in petioles without anastomosing bundles; and (2) that poisons stopped the movement, arguing that the living phloem cells should be put out of action by poisons which would not affect movements occurring in the non-living xylem. Czapek found that steaming of petioles and treatment with chloroform did prevent translocation out of the leaves; Deleano (1911) found that they did not. We need not concern ourselves at this distance of time with the reasons why their results contradicted one another for we have ample and conclusive evidence on the effect of inhibitors from later work and the question will be discussed in Chapter 12. Deleano's results, however, left the possibility open that the accepted view of translocation by the phloem might be wrong. Birch-Hirschfeld (1920) questioned it, and when Dixon came to make the sort of calculations discussed in Chapter 1 and to express the transfer of dry weight as flow of a sugar solution, and when he found speeds of the order of 40 cm hr^{-1} to be required for a hypothetical 10 per cent solution, he concluded that such movement was not possible in the phloem. He proposed instead that the outer zone of the wood was the channel of movement, and that the ringing experiments which seemed to show that cutting the phloem stopped migration really caused the blocking of those wood elements nearby that performed the transport.

During the 1920s opinion was sharply divided between those who followed Dixon in thinking the phloem area was inadequate to carry the measured masses and those who retained the classical view which was, after all, still quite well supported by experiment. Of this classical school the foremost proponent was Curtis (1920, 1923, 1925) who tried to refine the ringing experiments to obtain unequivocal evidence in favour of the phloem. They were never in fact, in his hands, refined to a point where the evidence silenced his opponents. Presence or absence of translocation across a ring was gauged by many criteria: the amount of growth that occurred beyond the ring, changing green of N-deficient leaves importing nitrogen, accumulation of salts and dry weight; but all were long-term and uncertain measures whose assessment left time for regeneration and other changes in the operated tissues. There remained, too, the objection that in removing the ring of phloem the xylem had been damaged. These experiments did, however, indicate that removal of the xylem had less effect on growth of a shoot, provided the supply of water was maintained, than did removal of the phloem. Curtis was concerned specifically with *upward* transport of carbohydrates, nitrogen and ash, while, as he says (1925): 'Dixon has very logically carried his criticism beyond those experiments which are directed to determine what tissues carry solutes upward, and includes also those experiments in which an attempt is made to determine what tissues carry foods downwards.' We can see now that Curtis was limiting the scope of his experiments unnecessarily, but can sympathize with his approach, since, if the xylem was not carrying solutes upwards with the stream of water it was much less likely to be carrying them against the stream.

The necessary refinement of this kind of experiment which finally settled the question was achieved by Mason and Maskell (1928a). Mason, whose experiment with yams was considered in Chapter 1, worked with Dixon in Dublin and started from Dixon's viewpoint, and the yam data are displayed as offering additional evidence of the impossibility of such high mass transfers by the phloem.* On moving to Trinidad he determined to use the very considerable labour force at his disposal to settle this question, and was joined in this work by Maskell whose careful statistical planning and assessments were to play such an important part in the work. What was new in this work was a clear recognition of the mutual involvement of transport channel and transport substance and of the importance of observing changes within hours, not days or weeks. The argument ran thus: carbohydrate is produced by photosynthesis in the leaves during the first hours of daylight; earlier experiments having shown that this carbohydrate has mostly disappeared from the leaves by the following dawn, it follows that a wave of carbohydrate must be

---

* In 1926 he started a hare that led translocation physiologists and anatomists a merry chase down the years. This was his observation (Mason, 1926) that the phloem of the vascular strands of *Dioscorea* appeared to be interrupted at the nodes by 'bast glomeruli', little knots of non-sieve-tube tissue. This was cited many times as evidence that phloem continuity was not universally necessary for translocation. The hare was finally caught and shown to be a dummy by Behnke (1968, and papers cited there). The glomeruli are composed of sieve elements indeed, though these are short, tortuous and difficult to identify except in the electron microscope.

passing through the transport tissues of petiole and stem during afternoon and night. Therefore by measuring the fluctuations of sugar content in different tissues of leaf, petiole and stem, one or more substances will be found to increase in the day and decrease at night. These will be the transport substances, and the tissue in which the main changes first appear will be the transport channel. Difficulties of interpretation of this line of reasoning might be expected if there turned out to be a multiplicity of moving substances and/or rapid interchange between tissues, difficulties which indeed bedevilled the same authors' later work on nitrogen transport, but for carbohydrate movement the result was triumphantly simple and clear. They elected to measure the changes in cotton plants of water, sucrose, hexoses and polysaccharides in the leaf, bark and wood. The soluble sugars were measured as the concentration in sap samples expressed from the frozen tissues, polysaccharides by the hydrolysis of dried and powdered material, and all these measurements were made at intervals of two hours for a 24-hour period (Series 1) and at three-hour intervals for a 36-hour period (Series 2). Careful statistical planning reduced the variation between plants and revealed the significances of any differences found.

A difficulty in the expression of the results is at once apparent. Sap concentrations are easily compared, but in comparing changes in the amounts of carbohydrate present in samples of leaf bark or wood the amount cannot be expressed per unit dry weight since the carbohydrate which is changing is a component of the dry weight. To express it per unit fresh weight is little better for this

depends on the belief that variation in the water content of a leaf is, relative to the total water present, very small indeed. It is certainly true that variation in the water content of a leaf is in general less than would be indicated by the changes in ratio of water to dry weight. Further, since normally the leaf tends to lose water while it gains carbohydrates during the day and to gain water while it loses carbohydrates at night, the percentage variation in fresh weight will in general be less than that of either dry weight or of water content. There is no evidence, however, to show how closely the fresh weight approximates to constancy, nor would the foregoing considerations apply to cases in which plants had been differentially treated in any way which might affect their moisture relations or their carbohydrate metabolism.

They accordingly expressed the changes as proportions of the 'residual dry weight', that is the dry weight less the total carbohydrates, expecting this to represent the stable ground structure of the tissue, and they show that this is the most satisfactory of the three bases for comparison.

The variations they found are illustrated in Fig. 2.1, whence it is plain that the hypothesised scheme is working: that the sugar concentration

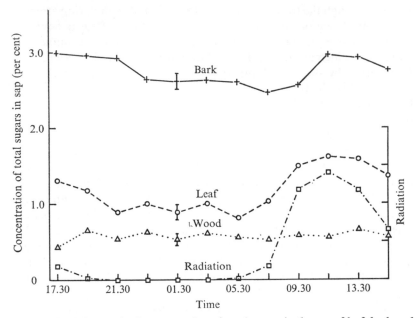

Fig. 2.1. The daily march of concentration of total sugars in the sap of leaf, bark and wood of cotton plants (g sugar per 100 ml sap). Radiation, as measured by the difference in ml between evaporation from black and white atmometers, is also plotted. Re-drawn from Mason and Maskell (1928*a*) Fig. 2.

of the leaf increases in the morning, and that it is the sugar content of the bark, not that of the wood, which shows parallel changes. Further, changes in the bark lag behind changes in the leaf. They were able to pinpoint more precisely what was happening by the careful use of correlation coefficients. The total sugars of the leaf sap, as can be seen from Fig. 2.1, are highly correlated with the total sugars of the bark if we take the partial correlation coefficients with time constant. The highest correlation coefficient was obtained between total leaf sugars at one sample collection and the sucrose concentration of the bark sap at the *next* sample collection.

This result is of great interest in that it indicates that changes in sugar concentration in the leaf may be reproduced within a period of a few hours in the bark at a minimum distance from the leaf in this experiment of about 50 cm. The evidence for associating these concentration changes in the bark with the transport of carbohydrate from the leaf is thus much stronger than if an even higher direct correlation between bark and leaf had been found, for the correspondence might then have been attributed to the effect of environmental changes acting independently, but in a similar manner on bark and leaf.

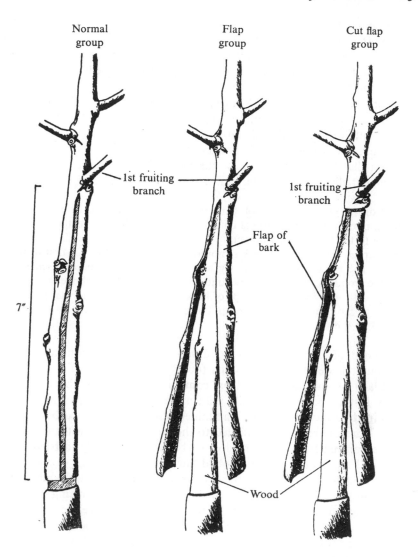

Fig. 2.2. The experiment which established the bark as the path of carbohydrate movement in cotton, Mason and Maskell's 'flap' experiment. Carbohydrate moved into the separated flaps of bark from above at undiminished rate, but the movement was stopped by the second ring in the cut flap group. Reproduced by permission from Mason and Maskell (1928*a*) Fig. 21.

Although this experiment established a strong probability that the transport was achieved by the movement of sucrose in the bark, it still left open the possibility of transport in the outer zone of the wood with leakage into the bark, even if the assumptions required were involved and improbable.

In order definitely to establish what this experiment strongly suggested, Mason and Maskell turned to ringing experiments, but with all the refinements of chemical and statistical analysis that had just been used in the observation of diurnal changes. The final and convincing demonstration that the phloem must be the path of transport came from the experiment illustrated in Fig. 2.2. There were three groups of plants:

1. Normal group. A ring of bark removed seven inches below the first fruiting branch and four vertical cuts made through the bark extending from the ring to the first fruiting branch.

2. Flap group. A ring of bark removed seven inches below the first fruiting branch. The bark between this ring and the first fruiting branch separated from the wood to form four flaps continuous with the bark above; the wood covered with vaseline and the flaps of bark replaced in their normal position.

3. Cut Flap group. As Flap group, but the flaps of bark were severed from the bark at the top.

The weight of carbohydrate in wood and bark of the regions above the ring was followed for 24 hours and found to change as in Fig. 2.3. The Cut Flaps of bark separated from contact with the bark above continuously lost carbohydrate, while the similar detached Flaps which were still joined to the bark above maintained almost as much carbohydrate as the Normal group still attached to the wood. In contrast, the wood, when separated from the bark, lost carbohydrate steadily in both Flap and Cut Flap groups. The carbohydrate entering the bark of the Flap group must have all come vertically through the upper end and it did so at a rate very little diminished by the operation of separating the bark from the wood. There remained still the question of where the sugar had travelled to get to the wood of the normal group, whether vertically in the wood or transversely from the bark, and Mason and Maskell were able to show that it must have been the latter.

While they were collecting and examining these data, Mason and Maskell noticed that there was always a gradient of sugar concentration in the bark in the direction of transport, in contrast to the wood where there was no gradient. They returned to this observation in their second paper, confirming and extending it as we shall discuss in the next chapter.

By now it was apparent that Dixon must be wrong and the classical theory broadly right, that the seeming impossibility of the measured rates

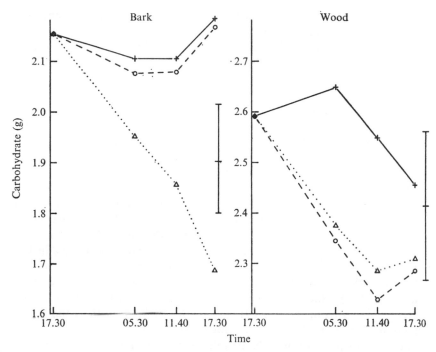

Fig. 2.3. The changes in the total carbohydrate during the cut-flap experiment in bark and wood of the 6½-in section of stem below the first fruiting-branch. Normal +——+; Flap ○----○; Cut flap △····△. Significant differences shown as vertical lines. Separating bark from wood has prevented translocation into the wood from above but not into the bark. Re-drawn from Mason and Maskell (1928a) Fig. 22.

of transfer had to be believed; and attention was swiftly turned to the question of how the transfer was made: what forces were operating on the sugar and how they were applied. We shall not follow this part of the investigation at the moment, beyond pausing to note that all explanations of the phenomena envisaged the process going on in the sieve elements of the phloem and that there was no experimental proof that the channel of movement was confined to these cells. Indeed, such proof is only just becoming available today. It is widely assumed, none the less, that the sieve elements are the channel, and it is necessary to examine the evidence.

First, there is the argument from shape. Sieve tubes, composed of files of sieve-tube elements, each elongated in the direction of transport, might be expected, by analogy with xylem vessels or tracheids, to be paths of longitudinal movement. This suggestion contains the implicit assumption that movement is more rapid within an element than between elements, tacitly supporting those mechanisms depending on flow of solution, for it is this kind of movement that would be faster within sieve elements than between them. Second is the argument from contents. The observation of Hartig (1860) already noted, that the sieve tubes could exude a strong nutrient solution, was not given its due weight until the 1930s when further work with phloem exudates was begun. These exudates will be considered in Chapter 8, when it will be clear that the contents of sieve tubes are just what would be expected of the translocation stream, namely a strong aqueous solution of sucrose whose concentration varies with the assimilatory activity of the leaves supplying translocate, containing also amino acids whose concentration may rise very considerably when the supplying leaves are exporting their organic nitrogen before abscission. Third is the observation of Mason and Maskell (1928*b*) on the changes of sugars in different radial fractions of the bark. The fluctuations of sucrose in the bark discussed above were occurring in the inner layers of the bark where most of the sieve tubes were found. 'It is difficult to resist the conclusion that the sieve tubes are the main channels of longitudinal transport, and that they are to a considerable extent isolated from the other tissues of the bark.' Fourth is the 'eosin reaction' of Schumacher (1930). Schumacher found that fluorescent dyes could be transported down the petioles of *Pelargonium* from an application on the lamina, and that if eosin was so transported it induced the formation of callus plugs on the sieve plates, and at the same time blocked the transport of other substances from the leaf. He argued that the transport must be going on in the sieve tubes since it was prevented by the callus formation and since the eosin had no other harmful effects; but it was never certain that there might not be other harmful effects. An exactly similar and less harmful blockage has been achieved by McNairn and Currier (1968) by the warming of a cotton hypocotyl to 40–45 °C for 15 min over a distance of 4 cm. Callus formed on the sieve plates; translocation stopped. In a few hours the callus was removed and translocation resumed. Fifth are the observations of Schumacher (1933), Bauer (1949), Eschrich (1953) and others that fluorescent dyes may be seen moving in the sieve tubes. It may be objected that these substances are not natural translocates, but the observation is highly suggestive. Sixth is an argument by default: that none of the other cells of the phloem seem at all suited to the function. Taking all these arguments in one pan of the balance, they seem to outweigh the objections that can be placed in the other pan: Dixon's, that the measured rates are impossible; and the impermeable appearance of the sieve plate. Certainly the sieve plate will offer high resistance to passive flow of solution, but unless we are certain that such flow is taking place the presence of sieve plates is not an argument against transport in the sieve tubes. It might be thought that a certain decision could be made with the help of radioactive tracer. We shall now examine what has been found by this new method and why techniques of sufficient precision are only currently being devised.

The image produced by radioactive emissions on a photographic plate first drew Becquerel's attention to radioactivity, and ever since, the ability to produce a visible image of the distribution of such substances (an autoradiograph) has been a powerful complement to their quantitative assay. The resolving power of such pictures, the capacity to distinguish as distinct images two very small radioactive sources very close together, has been the subject of continual refinement until it is now possible to resolve sources less than a tenth of a micron apart under ideal conditions. The conditions for maximum resolution are:

(1) The emitted radiation should be weak. If a very small image is to be formed on the film it is necessary that the emitted particles should have a very short range in the emulsion. It is fortunate that two isotopes of great biological importance, ^{3}H and ^{14}C, have weak emissions and form precise images.

(2) The source must be close to the film. Microtome sections of radioactive tissue can be coated with thin photographic emulsions, reducing the distance between emulsion and the surface sources to a fraction of a micron.

(3) The specimen must be thin so that sources of radioactivity are not superposed. For light microscopy and ^{14}C, sections of about 1 μm are desirable.

(4) For microscopic observation of the image and tissue together the film must be thin; otherwise one is looking at the tissue through a thick photographic image. Films are available with an emulsion 5 μm thick, and can be made thinner over small areas. The technique can be applied (with ^{3}H) to ultra-thin sections in the electron microscope. For an account of the methods used in practice, see Rogers (1967).

The technical problems of producing a reliable high-resolution autoradiograph of biological material are rarely in this part of the process, but occur in the fixation of the tissue, the procedure for preserving unchanged the distribution of radioactivity from the living tissue to the thin microtome section ready for autoradiography. In certain problems this is easy. If the radioactive atoms are incorporated into some stable substances in a stable structure, such as tritium in the nucleic acids of chromosomes, there is little difficulty in preparing a reliable and precise radiograph; but if the radioactive atoms are present in a water-soluble substance, the preservation of their true positions while the tissue is killed, dried and sectioned is much more difficult. It is just this problem which faces us in preparing an autoradiograph of the labelled sucrose in transit in the translocation system, and though we can do it well enough with a precision of 50 μm or so, the localisation of the sucrose image to single cells of the phloem presents many difficulties. At the gross tissue level the technique is easy. Plate 1 shows the resolution possible by merely cutting thick (1–2 mm) sections of living organs carrying labelled translocate, freezing on dry ice, drying under vacuum, and exposing the dried sections to X-ray film. By these means many experimental verifications have been made of the fact that the phloem is the channel of movement since Colwell (1942) published the first pictures of ^{32}P in *Cucurbita*. Comparison of these first pictures with later ones made by the micro-techniques of stripping film (Biddulph, 1956; Bachofen and Wanner, 1962) will show how seriously the resolution is limited by the difficulty of fixation, notwithstanding the application of freeze-drying and vacuum embedding.

The presence of sieve tubes isolated from the other usual components of phloem tissue in the parenchyma between vascular bundles of *Cucurbita*, the so-called commissural sieve tubes, has always offered the opportunity of radiography in an uncomplicated system. Webb and Gorham (1964) have achieved a radiograph of ^{14}C travelling in these sieve tubes (perhaps with their associated companion cells) thereby proving half the case: that sieve tubes do carry translocate. The other half, that other cells do not, is apparent in the pictures of Trip and Gorham (1967) and, more plainly still, of Schmitz and Willenbrink (1967).

The most hopeful line of advance seems to be the usual freeze-drying followed by vacuum embedding in resin, sectioning with glass knives at about 1 μm, and radiography of the still resin-impregnated sections. Leaving the resin in place helps to immobilise the sugar which has almost certainly moved during the removal of wax embedding medium in the older techniques. Even so, such pictures probably show not the position of the labelled translocate *in vivo* (presumably the lumen), but the position it migrates to on drying (presumably the walls), and may not be much further help in showing the path of translocate within a sieve element.

Just as radioactive tracers have confirmed Mason and Maskell's view of the channel of transport, so also have they confirmed that these authors were justified in believing that sucrose was the substance moving. It is vastly easier to demonstrate this now that the separation and characterisation of substances can be carried out by partition chromatography, and the demonstration that $^{14}CO_2$ photosynthesised into a leaf can be extracted from the phloem of the petiole as sucrose within an hour or so is reduced to the level of simplicity of a class experiment.

But the concept of a single transport substance has had to be revised slightly. Although Ziegler (1956) found no other sugar in the phloem exudates of many trees, and although radioactive translocate has been shown to be nearly all sucrose in diverse species (bracken, Hamilton and Canny, 1960; grapevine, Canny, 1960a; *Metasequoia*, Willenbrink and Kollmann, 1966), Zimmermann (1957b) has shown that a group of sucrose derivatives are translocated in *Fraxinus americana* and *Ulmus americana*. These are all alpha-galactosides of sucrose:

sucrose–galactose   = raffinose
sucrose–galactose–galactose   = stachyose
sucrose–galactose–galactose–galactose = verbascose.

Moreover, one or more of them often turns up in trace amounts in the phloem of other plants. Trip, Nelson and Krotkov (1962) fed these sugars labelled to two species that make them naturally, lilac and the white ash, and found them extensively interconverted in transit. There were exchange pools consisting of these four together with glucose, fructose, galactose, mannitol and sorbitol. Whatever was fed the sucrose family and mannitol were found at the sinks and along the path in amounts that varied with the distance. The significance of the spasmodic occurrence of these particular alpha-galactosides is still obscure; Zimmermann (1958a) suggests that the unloading mechanism from the transport channel to sink tissues in the white ash operates by the addition and subtraction of an alpha-galactose residue. Labelled glucose and fructose are found occasionally in moving labelled assimilate (Swanson and El Shishiny, 1958; Canny, 1960a) but generally after a long time of translocation and in nearly equal and relatively small amounts, making it probable that they are derived from sucrose either by side reactions in the plant or during extraction. Mannitol was found in considerable amounts at some seasons in the white ash (Zimmermann 1957b); serine and malic acid were found moving as labelled assimilate in *Soya* by Nelson, Clauss, Mortimer and Gorham (1961), sorbitol in apple (Webb and Burley, 1964), steroids in *Phaseolus* (Biddulph and Cory, 1965). In spite of these exceptions it remains a generalisation of great breadth that the transported substance is sucrose, so broad and constant indeed, that we must expect that either the loading or the unloading mechanism, or less likely the transport mechanism itself, is highly specialised to deal with this sugar.

We have spoken so far as though the xylem had quite lost all claims to be the path in which organic substances move, but in the certainty of the proofs given above, we must not lose sight of the fact that there are a number of not unimportant circumstances in which transport may be in the wood, and further investigation will no doubt reveal others. The most striking example is provided by the 'bleeding' sap that wells from the cut wood of grapevines, maples, birch and many other trees in spring, and of these the most spectacular is the bleeding from the Sugar Maple. Accounts of the conditions which influence this bleeding together with analyses of the saps may be found in Schroeder (1869–70), Jones, Edson and Morse (1903) and Priestley and Wormall (1925), which make it clear that the sugar content of this sap is generally less than 1 per cent. The sap from the Sugar Maple may be as strong as 2, or even up to 8 per cent sugar. When sugars and other solutes are present in the transpiration stream it is likely that they will find their way to the leaves and may play a significant part in the distribution of food reserves mobilised in spring, but except in the Sugar Maple this contribution is likely to be small compared with the phloem transport. Another possible example of xylem transport is provided by autoradiographs of labelled assimilate moving into fruits of *Phaseolus* (Bachofen and Wanner, 1962) where movement into the fruit seems localised to the newly-formed xylem. When the label was assimilated into the fruit and the rest of the plant darkened, reversing the normal gradient, outward transport from the fruit was in the phloem. So far as is known this is an isolated case, and most transport to fruits is stopped by operations interrupting the phloem. Moreover an attempt to explore the problem further gave the opposite result, movement only in the phloem (Othlinghaus, Schmitz and Willenbrink, 1968). As these authors point out, nodulated legumes are special in re-circulating the carbon skeletons from assimilate arriving in the roots rapidly back up the xylem as amino acids (Pate, 1962), from where it is built into protein all over the plant (Pate and O'Brien, 1968). It is quite likely that Bachofen and Wanner were detecting this stream. Strasburger's reservation (1891, p. 900) that the Umbelliferae were an exception is probably due to faulty technique (Münch, 1930, pp. 200–4). It is interesting that in these two classes of transport system when the sinks are vigorous and the need for solute transport great, that the usual phloem system may be supplemented by movement in the xylem. We shall be well advised to keep the possibility of some xylem transport in mind, while granting that in the majority of systems the movement is confined to the phloem.

Before leaving the topic of the transport channel, a word should be said about the names of the cells in the phloem. For the hundred years following Hartig's discovery of sieve-like wall structures, during the exploration of similar cells in many groups of plants by workers writing in German, French and English, a loose common vocabulary grew to be accepted for describing them. The usages in English were crystallised by Cheadle and Whitford (1941) and Esau (1950), to which last the reader is referred for a lucid account.

In summary:

sieve areas: pit-like recesses where pores with connecting strands are grouped.

sieve plate: a wall or part of a wall bearing one or more highly specialised sieve areas with particularly conspicuous connecting strands. If the pores are in clusters it is a compound sieve plate.

sieve tube: a file of cells disposed end to end with sieve plates on the end walls between them.

sieve-tube element (or member): one unit of a sieve tube. 'Thus a sieve tube member may be characterized as a cell in which certain sieve areas are more highly specialized than others, the former being largely localized on end walls to form the sieve plates.' Typical of angiosperms, and accompanied by companion cells.

sieve cell: an element where the sieve areas on all walls are equally specialised and no wall regions can be classed as sieve plates. Sieve cells have long tapering ends which overlap each other. They are not arranged in files like sieve-tube elements. Typical of gymnosperms and pteridophytes. Sieve cells have no companion cells, but in gymnosperms may be associated with special ray cells, the albuminous or Strasburger cells (see Kollmann, 1968).

sieve element: a comprehensive term which includes both sieve–tube elements and sieve cells.

In this book this terminology will be followed, though with some regrets for the loss of the hyphen in the non-adjectival 'sieve-tube'. Not only has it a respectable history; it represents the accent of pronouncing the compound word.

3. The diffusion analogy and the origin of the gradient

The meaning of analogy in logic is inference or procedure based on the presumption that things whose likeness in certain respects is known will also be found alike or should be treated as alike, in respects about which knowledge is limited to one of them. It is perhaps the basis of most human conclusions, its liability to error being compensated for by the frequency with which it is the only form of reasoning available.

Fowler

Diffusion is the statistical smoothing out of concentration differences due to random motions of molecules. If concentration differences exist in a closed system, that is there are more molecules of substance in one part of the system than another, the probability that the larger number will invade the space occupied by the smaller number is greater than the reverse; the system tends towards the most probable state, that of an even distribution of molecules. The description of this process is called Fick's Law, and that part of the law which concerns us first is the description of a steady state where a space has two boundaries, one where substance is maintained at a concentration C_1 and the other at a lower concentration C_2. Then the rate of transfer of substance through unit area of the boundaries and across unit distance of the space is proportional to $(C_1 - C_2)$. The constant of this proportionality, D, is called the coefficient of diffusion, and

$$\text{rate of transfer per unit area and distance} = D(C_1 - C_2). \quad (3.1)$$

Now putting in the area (A) of the boundaries, and x, their distance apart,

$$\text{rate of transfer} = \frac{D(C_1 - C_2)\,A}{x}. \quad (3.2)$$

The units in which we measure these things are

$$\text{g sec}^{-1} = \frac{D(\text{g cm}^{-3})\,\text{cm}^2}{\text{cm}}$$

whence it is clear that the units of D in the c.g.s. system must be $\text{cm}^2\,\text{sec}^{-1}$. In these units, the diffusion coefficient of dilute sucrose in water at $25\,^\circ\text{C}$ is 5.2×10^{-6}, and of molar sucrose at $18.5\,^\circ\text{C}$, 2.8×10^{-6}.

Now it is clear from a number of experiments (Hüber, Schmidt and Jahnel, 1937; Ziegler, 1956; Zimmermann, 1957b) that gradients of sucrose concentration exist in the phloem, that the phloem near sources contains more sucrose than the phloem near sinks and that translocation of sugar takes place down a gradient of concentration. Equally, it is quite certain that these gradients are too small by many thousandfold to bring about diffusive mass transfers of the measured size. The sugar solution exuding from cut phloem of *Fraxinus americana* (Zimmermann, 1957*b*) was most concentrated near the crown and the total molar concentration in summer declined downwards from a value of 0.535 M at 9 m high to 0.428 M at 1 m, a gradient of 0.013 M per metre. After leaf fall in autumn the total gradient disappeared and the gradients of two of the component sugars were reversed. The precise measurement of these gradients in phloem exudates is rendered difficult by the daily fluctuations in sugar content of the phloem that Mason and Maskell were the first to recognise. Examination of the curves in Fig. 3.1 will show that although the general trend of sugar concentration is clear and the exudate at 0.4 m from the ground is weaker than that at 2.5 m which is in turn weaker than that at 5.0 m, the daily march of changes at all levels imposed

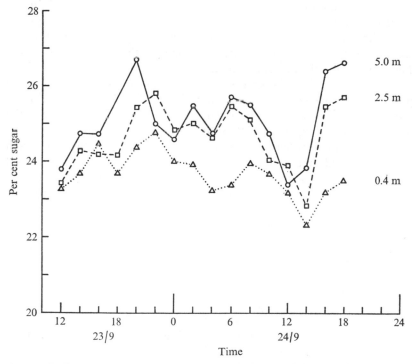

Fig. 3.1. Daily march of sugar concentration in exudates from three heights on the trunk of *Fraxinus americana*. Re-drawn from Ziegler (1956) Fig. 2.

on the innate sampling error makes the comparison at any instant of little value. The curves cross and recross and diverge. The curves found by Hüber *et al.* for *Quercus* in Fig. 8.4 are much better separated, and measurements from this figure yield values for the fall in concentration of from 3 to 4.5 per cent sugar over 11 m, or about 0.009 to 0.014 M sucrose per metre, close to the value found by Zimmermann in White Ash. Pfeiffer (1937) records averages of 0.026 M sucrose per metre for *Acer platanoides* and 0.01 M per metre in *Quercus rubra*. When the source of sugar is removed – a tree is defoliated – the gradient quickly disappears (Zimmermann, 1958*a*).

Furthermore, it was also clear from the solitary but extremely well-documented experiments of Mason and Maskell (1928*b*) that the rate of translocation was directly proportional to the magnitude of the gradient of sucrose in the phloem. It is to be regretted that the labour of collecting these results was so great that no other workers have been tempted to confirm them in other plants, but it is equally a matter for rejoicing that the one set of data is so unequivocal. Mason and Maskell pointed the analogy with the diffusive system and calculated the apparent diffusion coefficient of sucrose in the translocation system, showing it was $2-4 \times 10^4$ times that of simple diffusion. They noticed the general correlation of rapid movement with a steep sugar gradient in their first paper, where they showed that the movement of dry weight into the boll was four times greater by day, when the sucrose gradient was steeper, than by night; gradient and movement were significantly correlated. The result of ringing a stem, too, the rise in sugar content of the bark and leaves above the ring, required for explanation that the abnormal rise in sugar concentration of the bark retarded the rate of sugar transport from the leaf, which in turn required that the rate of transport out of the leaf should depend on the gradient of sugar from leaf to bark. The second paper is largely concerned with the detailed investigation of the analogy; to see how far the analogy could be pressed and, numerically, by how much the normal diffusion process was accelerated.

They tried to reverse the gradient and find the effect on movement, a technique which failed to produce import into darkened leaves, but which half succeeded in the stem, though the newly-established gradient was measured as sugar in the bark as a whole, not as sucrose in the sieve tubes. Movement occurred from the leaves at the base into a higher region of the bark but the declining gradient upwards was not statistically significant. The authors were fairly satisfied that closer examination would have revealed a reversal of the gradient of sucrose in the phloem down which the reversed movement was proceeding.

Constricting the area of a short length of the path of a diffusion system would, by diminishing the amount transported, be expected to steepen the gradient, giving a resulting transport which is more than that predicted from the mere proportion by which the path is reduced. Mason and Maskell show that if the area A is constricted to a for a short length l of its total length L, equation 3.2 becomes

$$\text{rate} = \frac{D(C_1 - C_2)A}{L - l + lA/a} \qquad (3.3)$$

and they tested the cotton bark for the closeness with which the translocation process followed this formula when either the area of the bark was restricted or the length of the restriction was varied. Considering the difficulties, the observed and predicted values are impressively congruent.

In both experiments the agreement between the predicted and observed rates of transport was generally satisfactory. In view of the difficulty of obtaining a true estimate of the gradient in the channels of transport, the correspondence must be considered remarkable. The results emphasize the close analogy between diffusion and the transport of carbohydrates in the stem.

Finally, Mason and Maskell were able to average their considerable mass of data and express the result as an apparent diffusion coefficient of sucrose (D in equation 3.2). For each collection they had values for a mean gradient in total sugars or sucrose, a time of translocation and an amount of carbohydrate moved. These, together with the cross-sectional area of sieve tubes, were used to calculate what we have called the specific mass transfer in Chapter 1 for a gradient of 1 per cent sugar per centimetre. The variation of this quantity from one sample to another was impressively small (the standard deviation of the mean of nine estimates was 11 per cent of the mean) as the reader may judge from Fig. 3.2. Here is shown a plot of rate of movement against gradient and the line of best fit whose slope gives a numerical value for the coefficient. The values of the gradients were much higher than those found in tree exudates, ranging up to 0.26 M per metre (total sugars). For calculation of the coefficient they used a specialised measure, the 'effective gradient', which allowed for the different diffusivities of sucrose and hexose in their total sugar fraction. The value of the coefficient calculated from their data was 6.97×10^{-2} cm^2 sec^{-1}, a value, as they pointed out, close to the coefficient for *gaseous* diffusion of a molecule the size of sucrose. The temperature at which the measurements were made was between 26.1 and 27.1 °C. To emphasise the difference of this coefficient from that of molecular diffusion (D), it will be given a different letter, K. The importance of the numerical value of K will become apparent in Part Two.

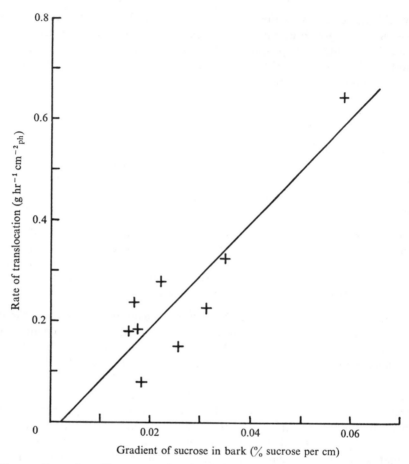

Fig. 3.2. Rate of specific mass transfer of carbohydrate in cotton (g hr^{-1} cm^{-2} phloem) plotted against gradient of sucrose in the bark (per cent sucrose per cm). Summary of the measurements made by Mason and Maskell with their fitted regression line from which they calculated a value for the translocation coefficient K. Re-drawn from Mason and Maskell (1928*b*) Fig. 14*b*.

Mason and Maskell were at a loss to explain their results in terms of a mechanism, and at this stage did not attempt to do so. But in the controversy that grew up in the 1930s between those who followed Curtis in believing the acceleration was brought about by circulation of streaming protoplasm in sieve elements, and those who followed Münch in thinking the motion was a mass flow of sugar solution through open sieve pores, Mason and Maskell found themselves obliged to make some decision on a mechanism. Reluctant to abandon the clear and notable fact of

Fig. 3.2 which did not fit the mass flow hypothesis, and yet having convinced themselves by calculation that circulation of streaming protoplasm would not supply sufficient acceleration and would be impossibly wasteful of energy, they proposed a third 'accelerated diffusion' hypothesis. No detailed explanation was given of how the acceleration might take place and the hypothesis remained untestable, indeed little more than a name expressing their faith in the importance of Fig. 3.2. That the fact has remained dormant ever since, contributing nothing to experimental planning and interpretation or to the elaboration of hypotheses, does not belittle it. The quality of the work and the certainty of the fact are unquestioned, and the ultimate theory of translocation must accommodate this proportionality of rate and gradient.

Suppose we assume that the proportionality applies to the mass transfer figures collected in Table 1.1, and that each may be compounded of a product of a gradient of sugar concentration with a 'diffusion' coefficient. Then we may explore the kind of sugar gradients that would be necessary to bring about these transfers if the 'diffusion' coefficients in all these different systems were similar to that measured in cotton. No matter that the coefficients will certainly be different, we are looking only for an order of magnitude. Also, having found the gradients for one coefficient, everything will be in the right units, and the gradients can be corrected by simple proportion when we know more nearly what the coefficient should have been. The first step is to convert the easily-visualised units of specific mass transfer in Table 1.1 (g dry wt hr^{-1} cm^{-2} phloem) into c.g.s. units compatible with K in cm^2 sec^{-1}. This is done in the second column of Table 3.1 by dividing by 3600. The next column gives the gradient necessary if the coefficient of diffusion is 7×10^{-2} cm^2 sec^{-1}, again in c.g.s. units, g cm^{-4}. This assumes that the two area bases, that through which the mass transfers occur, and that used by Mason and Maskell for the calculation of K, are the same, as can be verified by examination of what Mason and Maskell used as their area-base and called 'sieve-tubes' (quoted in Chapter 1). Further insight into what area was measured is afforded in a later paper (Mason, Maskell and Phillis, 1936a) when they come to use the coefficient in a calculation:

This apparent diffusion constant has been calculated from the observed rate of transport per sq cm of 'sieve tube groups' in the phloem, and the estimated vertical concentration gradient in this tissue complex. About half the total area is made up of phloem parenchyma cells, the remainder being sieve tubes and companion cells. The rate of transport per sq cm of actual sieve tube would therefore be more than twice as great as that used in the calculations. As, however, the vertical concentration gradient may, like the transport, be restricted mainly to the actual sieve tube track, it seems unjustifiable on this ground to increase our estimate of the apparent diffusion constant. This figure is therefore retained as an estimate of the order of acceleration required.

TABLE 3.1   *Calculation of the gradients necessary to bring about the specific mass transfers of selected measurements from Table 1.1*

Author and plant	Specific mass transfer g sec^{-1} cm^{-2} ph	*Gradient required if $K = 7 \times 10^{-2}$ g cm^{-4}	Gradient M sucrose m^{-1}
Dixon & Ball (1922)			
potato tuber	1.25×10^{-3}	1.79×10^{-2}	5.7
Crafts (1931a)			
pear trunk	$2.5 \ \times 10^{-4}$	3.58×10^{-3}	1.05
Tammes (1933, 1952)			
palm fruitstalk	2.75×10^{-2}	3.93×10^{-1}	1.15
Crafts (1931a)			
Phaseolus petiole	1.39×10^{-4}	1.99×10^{-3}	0.58

* Effective gradient in the sense of Mason & Maskell.

It seems clear that the area basis they used for calculating both rates and gradients and K is the area expressed in Table 1.1 as 'phloem'. Finally, that we may compare the gradients with those quoted above, a conversion is made into M sucrose per metre, by multiplying by $10^5/342$. The values obtained are much steeper than the gradients measured in tree exudates, but at the lower end extend, as is to be expected, into the range of those measured by Mason and Maskell in cotton bark. (The reason that the cotton gradients are not actually within the range of the others is Mason and Maskell's use of the concept of 'effective gradient' already noted.) The startling discrepancy between the gradients of Table 3.1 and the tree-trunk gradients focuses attention on the fact that a tall tree, with root sinks separated so widely from leaf sources, must, if the transport operates by a diffusion-type mechanism, be a special kind of system. Concentration at the source cannot well be much more than 30 per cent sugar, nor at the sink, less than 0 per cent. So the gradient can never be steeper than 30 per cent per trunk length. (For 10 m, 3 per cent per m, or 0.088 M sucrose per m.) Such a system would obviously require special modifications in the way of storage reservoirs along the path which could be filled at times of little sink activity and drawn on down steepened gradients over short distances at higher rates when the need arose. The rays and wood parenchyma seem admirably adapted to this purpose. The problem of the tree will be taken up again in Chapter 16, section 2.

Over the short distance between a growing fruit and the nearest leaf cluster a much steeper gradient would be possible which would permit a far higher rate of transport, though we must not rely on the figures to within a factor of five or so because of the uncertainty in K. The gradient in the palm inflorescence stalk is absurd, but then so is the rate of sugar movement. The source of sugar is very close to the cut surface (Tammes, 1933).

It is proposed to dignify the coefficient K with a name of its own. To call it 'the diffusion coefficient of sucrose in the translocation system' is not merely cumbersome, it is misleading to the extent it implies the movement is a diffusion movement. The similarity really ends with the formal mathematical relation; in magnitude the coefficient is unlike a diffusion coefficient and it would be confusing to call it one. Its basic relevance to the problem of translocation should now be plain, nor is it to be despised because we have so far only one measurement of it, and that obtained by great labour. More values and simpler ways of getting them will appear. It will be named hereafter simply 'the translocation coefficient'.

The gradient of sucrose in the phloem has its origin at the source of sucrose, and for leaf sources the matter is succinctly stated by Phillis and Mason (1933): 'When it is recollected that nearly every cell in the lamina is engaged in the manufacture of sugar, while only a very small proportion of those in the petiole serve conduction, it seems probable that the diurnal increase in the actual conducting tracts must be enormously greater than in the assimilating cells.' The details of the origin of the high concentration were considered by Mason and Maskell (1928 *a* & *b*) who found the sucrose concentration in the phloem of the leaf veins was greater than in the mesophyll, but that reducing sugars were lower in the phloem. So they were led to postulate that sugar export from the mesophyll took place as reducing sugars and that in the phloem condensation to sucrose occurred. Leakage back was checked by the impermeability to the disaccharide, and thus the head of sucrose concentration was produced in the leaf veins. The question was re-examined by Phillis and Mason (1933) who showed that the earlier extraction technique was unreliable; that reducing sugars showed no rapid fluctuations associated with changing transport from the leaf; and that it was the sucrose here, as in the stem, that varied in response to the daily march of assimilation, to darkening, and to ringing. They confirmed Mason and Maskell's result that the sugar concentration in the veins was greater than in the mesophyll and so were left to conclude that

sucrose is either accumulated by the vein against a gradient, or else there is a static component of some sort in the vein. This static component might be regional, that is to say, the gradient from mesophyll to phloem might be positive and transport might take place across a sheath region with a higher mean concentration of sucrose than either mesophyll or phloem... It will be clear that if the diurnal *fluctuations* in sucrose concentration in vein and petiole were to exceed those in the mesophyll, the presence of a static component would not be sufficient explanation of the negative gradient, and it would seem reasonable to infer the presence of a negative dynamic gradient in sucrose. Similarly, if transport were checked by ringing the stem and this led to a greater accumulation of sucrose in the conducting tracts than in the mesophyll, the accumulation of sucrose against a gradient by the vein, or some part of it, would be indicated.

Experiments showed that both these things happened, and that in the petiole, where dissection was possible, the changes went on mainly in the 'inner bark', rich in phloem.

The origin of the sucrose gradient was therefore somewhere along the path from the mesophyll cells to the phloem, and Phillis and Mason turned their attention to the structure of the leaf veins. Fischer (1885) had shown how the companion cells occupied an increasing proportion of the phloem as the order of vein branching increased, and Phillis and Mason found the same thing in cotton:

As the bundle-ends are approached the sieve-tubes gradually diminish in cross-sectional area and the companion cells are greatly enlarged, so that the sieve-tubes come to represent only a small proportion of the area occupied by the companion cells. In the leaf of the cotton plant sieve tubes are present as long as there is phloem, and undivided mother-cells, similar in all respects to the large companion cells, are present for some distance away from the bundle-ends. The phloem of the fine veins therefore consists of small sieve tubes with large companion cells, and also large undivided mother-cells that resemble companion cells. Fischer refers to both as *Transition* cells. Phloem parenchyma is not present. A typical cross section of a fine vein, with two small sieve tubes embedded in seven large cells with somewhat dense contents, is shown in [Fig. 3.3]. If, as we have suggested, the phloem of the fine veins is responsible for the accumulation of sucrose against a gradient, it seems reasonable to assume that the transition cells are the agents concerned. Sucrose might move across the mesophyll to the border parenchyma down a gradient, while the transition cells remove it from the border parenchyma against a gradient. After accumulation by the transition cells, sucrose would be liberated into the sieve-tubes and thence be distributed throughout the plant. Polar distribution of sucrose by the transition cells is suggested.

They were not able to dissect out the fine veins where most of the loading seemed to be going on, and so could not determine directly the sugar concentration there. By inference, however, with transport proceeding down a gradient of concentration, and with the sucrose concentration of the petiole about ten times that of the mesophyll, the concen-

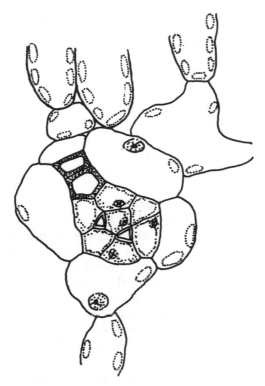

Fig. 3.3. Cross section of a vein of high branch-order in the leaf of cotton. Two small sieve tubes are embedded in seven 'transition cells', and all these, with two tracheids, are surrounded by the sheath of border parenchyma which abuts on the mesophyll cells. Re-drawn from Phillis and Mason (1933) Plate xxiii, Fig. 2.

tration in the fine veins further up the gradient must have been higher still. They were rather dismayed to find such clear evidence for another step in the translocation process which required non-physical explanation: 'the transport of carbohydrates is confronted by a further difficulty, namely the accumulation of sugar by the phloem against a concentration gradient. The physical difficulties here are quite as great as those presented by the enormous rates of movement achieved in the sieve-tubes.'

But we, with later knowledge of sugar-pumping mechanisms, in erythrocytes, in sugar cane parenchyma, and with clearer understanding of transfers of ions against electrochemical potential gradients, need not share their misgivings. This part of the process is easily credible. A similar polarised pumping mechanism that removed sugar from the mesophyll and loaded it into the veins was demonstrated in sugar beet leaves by

Leonard (1939). Once in the conducting system of a detached leaf, there was no tendency for the sugar to accumulate in any part of the petiole or midrib: the loading mechanism maintained a head of concentration at the source: and the sink having been removed from the diffusion-type system, the solute spread evenly in the channel.

Renewed interest is concentrated on these large parenchyma cells of the fine veins by the discovery (Gunning, Pate and Briarty, 1968) that some of them are the newly-discovered 'transfer cells'. The new histological procedures which are treated in more detail in Chapters 10 and 11, providing 1-μm sections that fully use the resolving power of the light microscope, have revealed a new cell type that is not recognised in either paraffin sections (because they are too thick) or in electron sections (because they are too thin). These cells have part or all of their inner wall surface elaborated into a labyrinth of fine projections (trabeculae, protuberances) whose effect is to increase enormously the surface area of the cell (Gunning and Pate, 1969a). Their name derives rather from their occurrence in places where much traffic must be going on across the boundary between apoplast and symplast, than from any direct experimental proof that they do transfer materials in specially large amounts. They are found in the pericycle of the leguminous nodule (Pate, Gunning and Briarty, 1969), in the placenta of ferns (Gunning and Pate, 1969b), in glands and nectaries, and frequently, though not always, the cells so far lumped together as transition cells in fine veins, can be distinguished as transfer cells of different types. Four types have been recognised (Pate and Gunning, 1969), and are illustrated in Plate 2.

Type A cells have ingrowths all round the wall and dense contents and are associated with the small sieve elements.

Type B cells are also in the phloem, but have the wall ingrowths localised to the wall shared with a sieve element.

Type C cells are associated with tracheary elements, have the ingrowths localised on the wall shared with the tracheary element, and less dense cytoplasmic contents.

Type D cells are cells of the bundle sheath with small localised ingrowths on the side of the xylem.

It is the cells of type A and B that are especially associated with sieve elements, and which must share with the transition cell/companion cell complex some special role in channelling assimilates from the mesophyll into the sieve tubes. The large surface area produced by the protuberances cannot always be a necessary part of the sugar-pumping machinery, since it is not by any means universally present.

Whether the sucrose pump is sited precisely in the companion (transition) cells as Phillis and Mason thought, in the sieve tubes themselves, or is spread over these and the border parenchyma and adjacent collecting cells, is a question that needs elucidation. Evidence that the pumping may occur in several successive stages comes from plasmolytic studies by Roeckl (1949). She was the first to use the elegant demonstration of the concentration jump that mesophyll cells (of *Robinia*) could be plasmolysed by sap exuding from the sieve tubes of the trunk, even though this sap was much further down the concentration gradient than the phloem sap of the leaf. With sections of leaves of *Ficus elastica*, where the structures were more easily seen, she investigated the variations in osmotic pressure of the cells by placing leaf sections in a series of solutions of increasing molarity. Solutions that just plasmolysed the palisade cells did not plasmolyse the collecting cells. To achieve their plasmolysis, the section had to be moved to a solution one or several steps stronger in the series. Moving the section back down the series, the collecting cells recovered from plasmolysis when the palisade cells were still plasmolysed. Roeckl says that the demonstration is so easy and clear that it could be used as a class exercise. On average, the collecting cells had an osmotic pressure about 20 per cent higher than the mesophyll. Regrettably the method was not extended to the further stages of the sugar path, the border parenchyma and the transition cells, but the speculation is not unreasonable that further steps will be found there on the rising ladder of sucrose concentration. It is only fair at the same time to record that Bauer (1953) failed to repeat Roeckl's observations and cites several authors who likewise found no gradient of osmotic pressure. This loading mechanism, an aspect of the translocation process that has received only the attention here outlined, seems a fruitful field for investigation.

4. Transport of substances other than sugars

It is manifest that many things other than sugar must have moved into the pumpkin we discussed in the first chapter to form the fruit, and though the major sugar movement was conveniently considered first, attention must now be given to the other substances that are transported about the plant body. Plants are organisms possessed of remarkable powers of synthesising complicated substances from simple ones, to the extent that most other living organisms employ elaborated substances prepared by plants, but how far this power resides in isolated organs rather than in the plant as a whole is a matter still debated. Much evidence suggests that there is considerable traffic of complex molecules between plant parts, and our growing pumpkin may have drawn upon amino acids, lipids, nucleosides, etc., synthesised elsewhere in the plant, rather than making them within itself from carbohydrate and imported mineral salts. To what size of molecule this traffic may extend we are yet ignorant: whether polypeptides, nucleotides, dextrins are imported intact, or even considerable pieces of organised protoplasm, mitochondria, plastids or fragments of nuclei. One thing is quite certain, giving the pumpkin all credit for synthetic ability it has still imported nitrogen from the rest of the plant and the minerals that are left behind as ash on burning. It was to the problems of transport of these substances that Mason and Maskell turned when they had so strikingly clarified the question of carbohydrate transport (Maskell and Mason 1929a & b; 1930a, b & c). They still had the tissue samples collected for the sugar work, dried bark, wood, and leaf sections with the nitrogenous substances present as these had responded to the ringing and other treatments. It seemed a simple matter to analyse the samples for different fractions of nitrogenous substance and unravel the familiar story of channel, transport substance and gradient, but they found a situation of such disheartening complexity that no one has since been bold enough to look at this aspect of the matter further.

Analyses of the plant samples for total nitrogen were in accord with the older view that organic nitrogenous compounds were formed in the leaves from nitrates arriving with the transpiration stream and exported from the leaves in the phloem. Movement of inorganic nitrogen into the leaves was unaffected by the removal of a ring of bark, but the same ring caused accumulation of nitrogen in both leaves and bark above; total nitrogen of the leaf increased by day and decreased at night; and, as with sugar, nitrogen was found to be translocated into flaps of bark detached save at their upper end. The rate of mass transfer of nitrogen (calculated as grams of asparagine per hour per square centimetre of phloem) was about one tenth of that of the carbohydrates (mean 0.03 g hr^{-1} cm^{-2} phloem). So far the process showed the same analogies with diffusion as sugar transport had done, and they set out to investigate the relation of movement to concentration gradients, and the 'acceleration of diffusion'.

The problem was a formidable one:

It will be seen that the solution of this problem involves the solution of a number of subsidiary problems, viz. (1) the channel of transport, whether the sieve tubes or some other tissue, (2) the nature of the mobile compound or group of compounds, (3) some estimate of the concentration gradient of the mobile form in the channel of transport. With the technique available hardly one of these questions could be settled independently of the others. The problem resembles that of the solution of a set of simultaneous equations.

Six separate fractions of nitrogenous substance were analysed in samples taken vertically in leaf, bark and wood, and transversely through the bark, but the results gave no support to their gradient hypothesis. Any gradient they could detect was in the wrong direction, getting more concentrated in the direction of transport, so they were led to postulate a declining gradient of mobile compound(s) in the sieve tubes masked by an increasing gradient of non-mobile nitrogen in other tissues of the bark. They examined the changes in gradients that accompanied changes in movement.

If we stop transport or reverse its direction we should affect only, or mainly, the dynamic gradient of translocatory material. While, therefore, the net gradient may be negative, the *change* in the net gradient, when the rate or direction of transport is altered, should be positively correlated with that change in movement, and should be a measure of the change in the dynamic gradient.

Further elaborate analyses and reasoning produced some support for the idea of a masked declining gradient of mobile nitrogen in the direction of transport, for reversing the direction of transport in the stem resulted in a steepening of the original increasing gradient. This, they argued, was due to the reversal of the masked declining gradient which was now added to the masking increasing gradient, not subtracted from it. In so far as quantitative estimates were possible the acceleration of nitrogen diffusion appeared similar to that of carbohydrates. 'This suggests that the acceleration mechanism in the sieve tubes acts impartially on sugars and nitrogen compounds'. The changes in the dynamic gradient were associated with the inner part of the bark where sieve tubes were most numerous. Of the nitrogen fractions analysed, they could not identify any one as wholly static or wholly mobile. Further attempts to identify the mobile compound(s) by observing changes in nitrogen fractions where the bark was interrupted by a ring led to a similar conclusion: that all nitrogen fractions *including the labile protein* contribute to the longitudinal movement in the sieve tubes. This conclusion was perhaps inevitable because of the fact that the different fractions proved very labile, easily converted from one to another within the bark more rapidly than they moved, but if the italicised words imply that protein was moving, a whole new vista of complexity is opened of which other glimpses may be gained from other observations, and which will take on new meaning when we come to the second part of the book.

Another set of studies of nitrogen movement was going on in Germany and was published while the Maskell and Mason series was coming out. Schumacher (1930) used as experimental system the loss of nitrogen by a darkened *Pelargonium* leaf to the rest of the plant which remained in the light. Total nitrogen was determined in samples of the lamina taken before and after a migration period of about a week and expressed per fresh weight of leaf, a measure that Schumacher showed was closely congruent with the expression per unit leaf area. This measure of trans-location was being used to determine the path of transport by a series of masterly surgical operations on the *Pelargonium* petiole. This petiole possesses a large central bundle from which Schumacher was able to dissect away the surrounding tissues, leaving and maintaining the lamina attached to the plant by a bridge of xylem and phloem only. The nitrogen movement still continued out of the darkened lamina. By technical skills difficult to imagine, he went on to refine the bridge still further until it contained either xylem or phloem only. In the one case the nitrogen movement ceased, in the other it persisted. The question of channel of movement has already been treated in Chapter 2, but the work is of interest in providing a second measure of nitrogen transport. In his Table 9 Schumacher lists three experimental determinations of nitrogen loss through phloem bridges whose area he records, and the relevant part is repeated here as Table 4.1. An obvious error (of the kind so easily made with metric units) in the unit of phloem area has been corrected. Schu-macher heads this column 'μ^2', which a little consideration will show to be absurd. From his statement (p. 792) that the phloem bridge in Ex-periment 42 (illustrated as his Fig. 6) had an area of '*ca* 0.13 mm^2' we see that this column should be headed mm$^2 \times 10^{-3}$. This is confirmed by

TABLE 4.1 *Values of the mass transfer of nitrogen across bridges of phloem.* (From Schumacher, 1930)

No. and dates of experiment	Total amount of protein N migrating during experimental period (N × 6.25) μg	Protein migrating per hour μg	Cross section of isolated phloem $mm^2 \times 10^{-3}$	Specific mass transfer g amino N hr^{-1} cm^{-2} phloem
42 15-22/4	29,230	174	128	0.136
51 19-26/4	26,000	155	80	0.194
76 18-24/5	32,720	227	95	0.239

his own calculations in the rest of Table 9 where speeds of flow of a 5 per cent protein solution are derived after the manner of our equation 1.1, and where the area necessary to get his answers is in fact that given here. The analysed loss of elementary nitrogen was converted to loss of amino acid or protein by multiplying it by 6.25, allowing the nitrogen to make up 16 per cent of the moving organic molecule. It will be noticed that the amino-nitrogen transfers in this system are from four to eight times those calculated by Maskell and Mason in cotton. Fischer (1936–7) used the same nitrogen movement out of darkened leaves to assess the effects of wilting, narcosis, etc., on translocation, but though he widened the range of species for which the movement could be shown, he made no phloem measurements and can provide no additional figures for mass transfer.

The work here briefly outlined constitutes very nearly all that is known of the movement of organic nitrogen in phloem. We known now that the exudates from sieve tubes contain amino acids (Mittler, 1953) and that the nature and quantities of these vary much with season (Ziegler, 1956), facts that fit well with the general picture drawn by Maskell and Mason. This picture receives its complement from recent Russian work (Kursanov, 1957) which has contributed a concept of circulation of carbon atoms from leaf assimilates to the roots in the form of sucrose; transformation there of labelled carbon into amino acids, part of which is used for synthesis of root material and part returns to the metabolic sinks of the shoot still in the form of amino acids. Similar work by Pate on nodulated legumes was referred to in Chapter 2. But the lack of any radioactive isotope of nitrogen has severely limited the development of this knowledge in the general directions that our understanding of carbon transport has broadened. Chromatography of labelled nitrogen moving in the phloem would tell us much, as perhaps would radiography of the labelled tissues. The isotopic substitutes for radio-N: ^{15}N with mass spectroscopic analysis, and ^{14}C-labelled amino acids, are both unsatisfactory and of formidable complexity. The former has never to my knowledge been attempted in translocation studies, probably because the sample sizes available are so small; the latter provides little hope of a useful return for much work since the label can very easily become detached from the nitrogen and is likely to turn up in a multiplicity of non-nitrogenous compounds of no relevance. Moreover, the introduction of nitrogen label into the phloem can be achieved by no major ready-made natural pathway like the photosynthesis of labelled carbon.

The demonstration by Russian workers of amino acids translocated from root to shoot, added to Mason and Maskell's demonstration of nitrogen movement out of the leaves, raises a suspicion that we may have been blinded by the simplicity of the unique compound of assimilate translocation, sucrose, to a whole class of other transport substances that do not exchange carbon with the products of assimilation in leaf and stem and so do not become quickly labelled. Sucrose being a relatively stable compound, it is possible to picture many other substances moving in the sieve tubes alongside labelled sucrose, but retaining all their carbon atoms unlabelled, and so not turning up in our radio-assays. We are certainly examining the stream of freshly-made assimilate by tracer techniques, but this is surely only one of the streams in the sieve tubes. The two amino acid movements, out of mature leaves and away from the roots, must be isolated either spatially or chemically from the assimilate stream. The difficulties of getting suitable label into these other streams fall in the context of the introduction of foreign substances to the plant body to be discussed later in this chapter.

Looking beyond movement of carbohydrate and nitrogen, there remain of native substances the class of inorganic mineral atoms, and a multiplicity of smaller classes that might be moving or might be synthesised as required. For the reason that these latter are present in such small amounts, unless some easy and specific assay was available they have been left uninvestigated, and often their presence in phloem exudates is the only suggestion that they might be translocated. Auxin can be detected by bioassay with certainty in minute amounts and has been shown to be present in phloem exudates (Hüber *et al.*, 1937) and to move there (Eschrich, 1968), but it is certain that there is another kind of transport system for the growth substance operating in the small herbaceous stem and root systems most studied, and one which has defied many attempts at decipherment. Enzymes involved in sugar phosphorylations and especially phosphatase were found in phloem exudate by Wanner (1953*b*). Peel and Weatherley (1959) record the presence in exudates of organic acids, particularly citric and tartaric, and of potassium ions as the balancing cation. Further investigation may show that all these things are moving, but being chemically isolated from the assimilate stream and forming anyway a small fraction of the total dry-weight moved, they have been neglected.

Taking now briefly the category of mineral ions, it is necessary to divide the evidence into those experiments made on ions naturally present in the plant and those employing tracer ions, which, when introduced by incisions, must be considered as foreign substances. Because incisions are nearly always necessary to introduce radioactive ions anywhere but in the roots, the introduced ion may get into a system where it would not naturally be found, and, spreading there, give a false clue to the natural transport of the ion. For the movement of native mineral ions we return to further work of Mason and Maskell (1931, 1934) where they first analysed their dried cotton samples from the carbohydrate and nitrogen work for phosphorus, potassium and calcium, and later prepared a new set of samples to study the changes in gradients at different ages. The same kind of evidence as had been obtained for nitrogen suggested that potassium, phosphorus and other ash constituents ascended the stem in the wood and were re-exported to the roots by the phloem. Calcium also ascended in the wood but did not return from the leaves. The relation of gradient to movement was again not as clear as had been found for carbo-hydrate and the authors once more suggested a masking gradient of static material. Radioactive ions introduced via the roots should give unexceptionable evidence of the ion distribution and a number of such experiments were done very soon after tracers became available. The most refined and technically satisfactory were by Stout and Hoagland (1939) who supplied radioactive phosphate, sodium, potassium and bromide ions to the roots of cotton, willow and geranium in sand and water culture and found that regardless of plant or ion, the minerals spread rapidly in the transpiration current to all tissues. Separating bark from wood with a piece of oiled paper, they found no vertical movement into the bark from either end, and concluded that movement of the ions was normally in the xylem not the phloem, but that equilibration across the cambium was very rapid; ions found in the phloem had arrived there from the adjacent xylem.

This transfer from xylem to phloem or the reverse has been confirmed by other workers and must be constantly remembered in assessing the results of experiments with ionic tracers, particularly phosphate. Such transfer is clear in the experiments of Biddulph and Markle (1944) who managed to introduce labelled phosphorus through cuts in leaves without injecting the xylem, and found then that it moved in the phloem but leaked to the xylem and the leakage could be prevented by a layer of oiled paper. Phloem-limited movement was also established by radiography for sul-phate (Biddulph, 1956), and strong evidence that labelled caesium and potassium moving from bean leaves travelled by the sugar path (Swanson and Whitney, 1953) came from the almost total inhibition of the movement by a local low temperature, though the channel was not checked by radiography. Phosphorus in these experiments leaked to the xylem and its movement was little affected by the low temperature. This practice of introducing mineral tracers through cuts in the leaves, though tempting when penetration through the intact cuticle is slow, can lead so easily to the injection of the xylem vessels following the release of hydrostatic tension, that it should be avoided if possible. Colwell (1942) recognised this danger in his early studies with radio-phosphate and showed how the plant must be kept fully turgid if phosphate entering through cuts is to be confined in the phloem.

Calcium occupies a special position among the studied ions in being non-translocatable by the phloem. This divalent cation is used by plants in a number of structural and co-factor situations and has profound effects on protoplasm, cell walls, and ecological distributions. Drastic effects are produced by too little or too much of it, and the tolerances and requirements for it vary over a wide range. As the oxalate it is present as crystals in many cells, but whether this is a means of removing calcium or oxalate from systems where it is not wanted, or whether the crystals play some more positive part, is not clear. Notwithstanding all this traffic and importance it seems that the phloem cannot carry it. It is absent from all the analyses of phloem sap except those made by Moose (Table 8.1), which makes these estimates a little doubtful. However applied: to leaf surface, by cut flaps, by root feeding or to cut shoots, it seems to spread exclusively in the xylem. Once it gets into leaves, it stays there and falls with them. It is not re-circulated, as potassium can be (Greenway and Pitman, 1965). Its spread is unaffected by steam rings or cuts in the bark or other treatments that stop phloem translocation in contrast to magnesium which moves like other ions. Being divalent, it exchanges rapidly onto ionic binding sites of the xylem, and once there is difficult to dislodge. It can be found in the phloem, but seems always to arrive there indirectly from the xylem, not by transport in the sieve tubes. These statements may be followed up in the papers by Hoad and Peel (1965), Shear and Faust (1970), Steucek and Koontz (1970). A differing view that speaks for some phloem mobility of calcium is put by Ringoet, Sauer and Gielink (1968), but most workers accept the inability of sieve tubes to carry it.

Similar difficulties of introduction are present in testing truly foreign substances for transportability in the plant: that it is difficult to get them in without upsetting the system. Fluorescent dyes were at one time a favourite tool and in Schumacher's hands (1933, 1937) provided much valuable information. The less toxic, such as fluorescein, could be seen moving in sieve tubes apparently taking advantage of the sugar transport mechanism. Attempts to study this movement on short sections of plant organ (Bauer, 1953) gave conflicting results and low rates of transport. This is the first occasion we have had to notice what is a constant property of the translocation system: it does not work if you cut it up. We shall constantly be meeting examples of this extreme fragility and may notice here that it is an extra hazard in getting foreign substances in. Fluorescein was applied by W. Schumacher (1933, 1937) in a gelatine drop to the scraped nerve of a *Pelargonium* leaf, diffused through the cells between the surface and the phloem, and later spread in the phloem at rates of 2 to 30 cm hr^{-1}. The movement seemed to be independent of the native solutes for the speed was undiminished out of leaves darkened for as long as seven days, by which time carbohydrate transport must have been negligible, and nitrogen export proceeding rapidly. The fluorescein apparently moved down its own concentration gradient. Similar results were obtained by A. Schumacher (1948) in *Bryonia dioica*.

Bauer (1949) lists several other fluorescent dyes he found to move in sieve tubes: berberin sulphate, primulin, rhodamin B, rhodamin 6G and chinin chloride; and others that did not move. One would expect that a very great deal should be learned by studying fluorescent dye spreading in the sieve tubes but the difficulties of making the observations have so far limited this line of investigation. The conditions for translocation and those for good microscopical images are directly antithetical: for the one, intact plants, undamaged phloem and a minimum of handling; for the other, as thin a layer of cells as possible separated from neighbouring tissues. The compromises that have been achieved have been unsatisfactory both as translocation systems and as microscopic images: a thick section of phloem viewed of necessity at fairly low magnification. Bauer's description of what he saw as the front of fluorescein advanced along a sieve tube in a bridge of phloem cut in a petiole of *Bryonia* is perhaps the clearest and most reliable:

After about half an hour the dye-front appears as a diffuse veil in the field of view filling the whole width of the cell. At no time was the fluorescence of the side walls stronger than that of the lumen; rather, the fluorescence faded somewhat towards the sides. Some minutes later, when the dye-front had passed out of the field of view, a few starch grains in the parietal protoplasm began to fluoresce.

(Applied dye concentration, 1/1000; magnification × 200.) And with berberin sulphate:

Just as with potassium fluorescein, the dye-front moved forwards in the sieve tubes with a dull, diffuse fluorescence. Again the whole cell width seemed filled with the colour; because of the thinness of the parietal protoplasm layer it could not be distinguished whether the protoplasm was included in the fluorescing material. Only when the dye had reached the next or even several following cells did the first perceptible accumulation begin in the protoplasm, which now rapidly took it up. The dye was bound in fine reticular-granular structures. Moreover, as long as the dye was passing, there was visible through the glowing network of the protoplasm, a diffuse glimmer of fluorescence that could not have arisen from the underlying tissue because initially only one sieve tube carried the dye.

Similar effects were observed in petiolar phloem of *Pelargonium*.

Another foreign molecule whose movement has been much studied is the growth-regulator and weed-killer, 2,4-dichloro-phenoxyacetic acid (2,4-D), a short discussion of whose movement must serve as a general example of the way many similar artificial growth substances move. By virtue of the local bendings and other growth responses induced by small amounts of 2,4-D, its arrival at some part of the plant can be gauged, or, on extraction from the tissue, a bioassay will give a specific and quantitative estimate of its presence. Alternatively the molecule is available labelled with ^{14}C, but since it is liable to breakdown in the plant, mere presence of label cannot be taken as evidence of the arrival of 2,4-D. Applied to a leaf with suitable surfactants to aid penetration, it is found to spread in the phloem to apices, young leaves and roots, but not into mature leaves. Not only does 2,4-D move in the carbohydrate system, the evidence suggests that the movement depends in some way on a supply of sugar moving in the same direction. Transport of 2,4-D is faster from lighted or recently illuminated leaves than from leaves kept in the dark, and in the dark can be increased by the simultaneous application of sugar (Weintraub and Brown, 1950). Details of these experiments can be found in Crafts (1951) and Crafts and Crisp (1971).

In one of the neatest examples of co-operative evolution the aphids and viruses have evolved a relationship with the sugar transport system of plants, the aphids using it as a source of food, and the viruses, as a pathway of invasion of the plant body. Transmitted from one plant to another in aphids, and penetrating to and from the sieve tubes through the aphid mouthparts, a number of mosaic and phloem-limited viruses have occupied a tight and fascinating ecological position whose complexities have been discussed by Esau (1961). Once inside the phloem the virus particles move much as we have seen other substances, native and foreign, to move in the phloem, and being detectable by sensitive and specific assays, provide another class of substance whose movement is readily followed. A comprehensive account of the literature concerning virus movement may be found in Bennett (1956), from which a few general rules emerge: that viruses move in the same general pattern as the sugars, from sources of carbohydrate to sinks, and show the same reluctance to move in the reverse direction. On the other hand, viruses do not show quite the close correlation with sugar movement recorded above for 2,4-D; viruses move out of darkened mature leaves as fast as from illuminated ones. The fact that high concentrations of virus are found in the exudates from cut phloem of infected plants lends support to the idea that the process is occurring in the sieve tubes, and electron microscope studies reveal virus particles there (e.g. Esau, Cronshaw and Hoefert, 1967).

Even if we follow Zech (1952) in thinking that the virus transport is transport of small precursor particles rather than the fully-developed virus molecules, we have arrived here at a wholly new order of complexity in the transportable substances. So far the substances we have found to be moving have had molecular weights of at most a few hundreds, but if viruses can be translocated, we are justified in asking to what higher orders of size and complexity the phenomenon may extend. The answer is that we do not know. Apart from one highly interesting and suggestive paper there is no evidence for the translocation of highly-organised pieces of protoplasm, plastids or other particles. This remarkable paper by Lou, Wu, Chang and Shao (1957) starts from the observation, verifiable by any student, that the nuclei in the cells of onion bulb-scale epidermis occasionally migrate from one cell to another through pits. Consequently some of the epidermal cells contain two nuclei, some none, and now and then a nucleus may be distinguished in transit. The authors go on to explore the extent of this unexpected traffic, and, staining with nuclear stains, show evidence of nuclear fragments, among other stations, travelling in sieve tubes. We may therefore keep our minds open to the possibility of organised nuclear material and other large fragments of protoplasm being translocated.

The generalisation that crystallises from the diverse information listed in this chapter may be expressed in the form of a paradox: that while the translocation system seems to exercise the tightest control over the form in which sugar is transported there is no specific limitation of other substances or class of substance. It seems that almost anything that can be got into the sieve tubes without damaging them, and which is not itself toxic to the sieve tubes, can be translocated. Viruses and nuclear fragments, dyes and weed-killers, inorganic ions and organic molecules seem to be translocated in the same channel, in the same patterns, and at roughly the same linear speeds as sucrose. I suggest that this paradox implies that the transport mechanism itself is quite non-specific, but that the specificity for sugar resides in the loading, and perhaps also the unloading mechanism between other tissues and the sieve tubes. It is this loading mechanism which establishes the initial gradient of concentration of sugar (which we have seen is the factor controlling both transfer rate and direction) and, being most probably a typical active transport mechanism across cell membranes, is most likely to be highly selective of the molecules it can load. Here may lie the significance of the alpha-galactoside series of transported sugars: that the loading mechanism deals with sucrose and is little concerned if some other residue is tacked on to the molecule. The mass transfers of all the other substances are very much less and possibly limited by the rate at which they get into the sieve tubes, not by the transport capacity of the mechanism. The integrity of the amino acids, organic acids, etc., moving alongside sucrose in the sieve tubes seems to discountenance an alternative view that the sieve tubes contain a sucrose-synthesising system of great potency which tends to turn everything there into sucrose.

5. Patterns of movement

Was mir aus allen diesen Versuchen, die ich in grösserer Zahl angestellt habe, im einzelnen hier aber nicht beschreiben möchte, wesentlich erscheint, das ist das geradezu erstaunliche Regulationsvermögen der Pflanze, das in diesem Ausmasse bisher kaum geahnt werden konnte. In vorläufig noch ganz geheimnisvoller Weise ziehen diese Stoffströme, Bahnen öffnen und schliessen sich, Richtungen wechseln, Stillstand und heftige Bewegung lösen einander in buntem Wechsel ab.

Schumacher (1933)

The traffic of substance goes from source to sink, but which sources, we are justified in asking, supply which sinks? There are several different kinds of source and, of some kinds, a host of individuals in a plant body; each leaf is a separate one; each segment of each woody branch may act as a separate source; and there are many of a more ephemeral kind that function at particular periods, seed endosperms or cotyledons, and the variously modified storage organs of tubers, bulbs and corms. Correspondingly there are a variety and multiplicity of sinks actual and potential: shoot and root apical meristems, expanding leaves, cambia, fruits, tendril tips and (when laying down stores) all those storage organs and tissues already noted that can become sources at other seasons. Nor must the title of source or sink be applied sweepingly to any region without specifying what substance is being considered, since there may well be centres that are at the one time sources of one substance and sinks of another, as apices consuming carbohydrate and nitrogen yet exercise hormonal and other controls by substances they export. Considering this complexity we might despair of finding any simple answer to our question, but that, on investigation, a number of strict and simple generalisations emerge that can be comprehended in a single rule or law.

Such investigation was impossible while the only means available were those of classical chemical analysis. All sucrose was alike; it was not possible to distinguish what had been synthesised on the spot from what had migrated from elsewhere, nor what had come from one source from what had come from another. Attempts to unravel the puzzle by cutting the channels and interrupting the supply from selected sources were inconclusive and always open to criticisms that the operations had upset the patterns of flow, an objection whose reality has been fully justified by modern work. The necessary distinction between molecules of different origins became possible with the advent of isotopes, and our whole knowledge of translocation patterns is derived from experiments with labelled molecules.

There is perhaps no translocation experiment so rewarding, in the sense of a maximum return for the minimum effort, as the whole-plant autoradiograph which gives a visual display of the amounts and patterns of the movement of a labelled atom, a picture of the plant body with each part shaded in proportion to the amount of labelled translocate it has obtained. At the gross level of distribution into organs the technique is simple in the extreme. After the desired period of translocation of, say, labelled photosynthate from a particular leaf, the plant may be dried by pressing between sheets of blotting paper and, after mounting on clean paper, pressed to a sheet of X-ray film for the necessary exposure period. The drying may be hastened in an oven. Several refinements in the drying – the fixation of label in the tissue – may be necessary to prevent large-scale migration of the radioactive molecules as some parts dry more quickly than others. Of these the simplest is to dissect the plant into separate leaves, internodes, etc., when there can be no migration between pieces on drying, but the distribution within each piece must not be trusted. A radiograph prepared in this way is shown in Plate 3, and displays at once the paths taken by the labelled photosynthate from the source leaf and the amount of the transport to each sink, information whose collection by assay for radioactivity of the various plant parts would be tedious. There are, of course, reservations to be made about the radiograph picture. In so far as the density of dried tissue overlying the radioactive atoms varies from one part of the plant to another, differences in density of the image may not wholly reflect differences of radioactive content. Again, the presence of radioactive atoms in a plant part must not be taken as evidence, without further chemical investigation, of the presence of a particular labelled molecule that has been applied. In Plate 3 we know from other work that the labelled translocate is sucrose, but other applied molecules may be metabolised in transit or on arrival, and, over long periods, the ^{14}C of sucrose will be converted into other compounds. Care must be taken too about stray $^{14}CO_2$ either escaping from the vessel in which it is applied via intercellular spaces of the leaf, or released by respiratory breakdown of labelled photosynthate in the leaf and elsewhere. Thrower (1962) has shown that some labelled carbon detected by radiographs must be ascribed to fixation of labelled gas arriving by diffusion outside the plant since an untreated plant nearby gave a radiograph, and we shall have occasion to consider later the misleading effects of $^{14}CO_2$ diffusion within the gas spaces of the plant. But the advantages far outweigh the disadvantages, and as a preliminary survey of an unfamiliar translocation system the method is unrivalled.

In refining the method to give more precise localisation of radioactive material within plant organs, limits are set by the problems of fixation already noted in discussing microradiographs of radioactive translocate in the phloem. Nelson and Krotkov (1962) have shown that entirely different pictures are obtained from radiographs prepared by freezing source leaves and drying at −20 °C, from those obtained if the ice melts during drying. In the one case the radioactivity is confined in the mesophyll islands between veins; in the other it is mostly in the veins, the condition seen in the plants dried at room temperature, and which must therefore be regarded as an artifact. The labour and equipment required to perform this kind of freeze-drying fixation are so much more than in the simple press-drying fixation as to make the method unattractive as a routine survey, and unless high-resolution pictures are desired nothing is gained by it. Many workers prefer to achieve the less reliable picture with less work. Attempts to make quantitative estimates of the radioactive content of the dried plant parts by scanning with a detector are seldom satisfactory for the obvious geometric reasons; the plant parts should be converted to barium carbonate or ground to powder (O'Brien and Wardlaw, 1961) and assayed under standard conditions. Chemical analysis to test the integrity of translocated molecules should not be attempted on the dried parts since most molecules will have suffered some change by being dried in the tissue. Parallel experiments with suitable solvent extractions are usually necessary.

The patterns of movement that these experiments disclose have several clearly-defined characteristics expressible as generalisations. *Rule 1*, the movement of tracer shows a tendency to strict longitudinal confinement. Tracer moving from a single leaf is naturally at first confined in the stem to the phloem of those vascular bundles that form the leaf traces, but this restriction may persist over considerable distances, apparently unaffected by the anastomoses of phloem with other bundles. Tracer is found only along a narrow line of stem above and below the source leaf. Such confinement was shown by Hamilton and Canny (1960) in bracken, and by Thrower (1962) in *Soya*, and has been commonly observed in the other plants studied. Indeed, in the bracken, label occupied only a part of the ring of phloem surrounding a vascular bundle, and travelled many centimetres without spreading sideways into the adjacent sieve cells, though such spread might be thought to be facilitated by the wealth of lateral sieve areas. Mere contact of phloem strands in an anatomical picture seems not to imply that exchange of translocate is necessarily possible. This generalisation had already been deduced by Caldwell (1930) without benefit of tracers, in as elegant a piece of observation as one could wish for. The south side of a Swede turnip, he noted, was larger, and when assayed had more dry matter, than the north side. He measured many from rows planted both north–south and east–

west and recorded the radii along each cardinal line. The radii in the south and west directions were greater than those to the north and east, the differences being most marked in north–south rows. Let him tell it in his own words:

As it is clear that the only part of the normal plant in which carbohydrates are elaborated is the foliage, one must assume, from these figures, that the available amount of sugar and of other materials is greater in the leaves of the south and west sides. The materials manufactured in these leaves pass down to the south and west portions of the bulb. One must, further, assume that translocation across the bulb is very limited and that the difference in the amounts of food-material manufactured in the leaves which receive the greater amount of light is sufficient to cause structural differences between the various sides of the swede bulb. This idea is considerably strengthened by a comparison of the figures for the swedes from the N.–S. and the E.–W. rows. In the case of swedes from the E.–W. rows, the differences in the lengths of the radii on the S. and W. sides as compared with those on the N. and E. sides are greater than they are in swedes from the N.–S. rows. This is probably accounted for by the fact that the leaves of the latter swedes, as they increase in size, will tend to overlap the southern leaves of the adjoining plants on the north side and so to decrease the amount of material manufactured in these. The greater development of the west side of the bulbs is presumably associated with the fact that, in general, bright sunshine and high air-temperature most frequently occur together in the afternoon. This greater local development and deposition of carbohydrate can only be satisfactorily accounted for by the supposition that the greater opportunity for the manufacture of carbohydrates in the best illuminated leaves is associated with an absence of an equalising transverse translocation in the bulb.

Zimmermann (1960) tried to measure the extent of this confinement which he found in a tree trunk by defoliating one side of a Y-shaped tree and observing the effect on sugar concentrations in exudates collected round the trunk.

The defoliated side of the tree is very sharply defined. 6 out of 16 samples at 8.45 m, but only 3 out of 16 samples at 1 m height show the defoliation charac-teristics. Clearly, assimilates from the leafy side of the tree have spread out somewhat. What is the degree of this lateral conduction? If we assume the average circumference of the tree to be 50 cm then 19 cm of phloem width at 8.45 m show defoliation and 31 cm leafy sugar concentrations. The corre-sponding values for 1 m height are 9.5 and 40.5 cm respectively. The leafy side has therefore spread laterally $40.5 - 31 = 9.5$ cm over a length of 745 cm. The angle covered by such an arc is only 44 minutes!

We must conclude that movement is very much easier between con-secutive sieve elements in a file than between adjacent sieve elements in different files, and that lateral sieve areas have less to do with transport than sieve plates on end walls.

Rule 2, mature leaves do not import tracer. Plate 3 shows this clearly: no radioactivity is detectable in the mature leaves. This is another aspect of the failure that Mason and Maskell recorded in their attempts to reverse the gradient of sugar into mature leaves and induce a net import of dry weight. The generalisation carries a corollary: since very young leaves must grow at the expense of food imported from elsewhere, there will be a change during the development of a leaf from its being purely a sink to being purely a source. The change is apparent in the work of Jones, Martin and Porter (1959) on tobacco, and has been further investigated by Thrower (1962) in *Soya* leaves, where the expanding leaf imports labelled photosynthate from the neighbouring expanded leaves until it reaches 50 per cent of its full size. Maximum import occurred at about 30 per cent expansion. The rising capacity to export label overlapped slightly the declining import, starting at about 30 per cent expansion and becoming rapidly greater above 50 per cent (Fig. 5.1). A reservation must, however, be made about this generalisation; mature leaves do

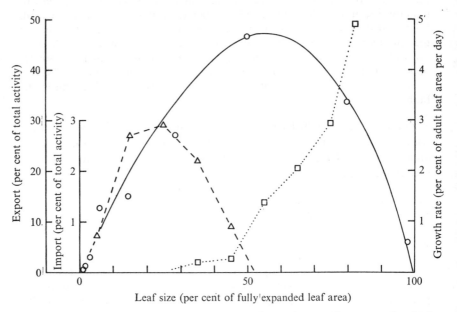

Fig. 5.1. Comparison between import, export and growth rate of an expanding leaf of *Soya*. Growth rate is measured by the daily increase in percentage of adult leaf area. Import △----△; export □····□; growth ○——○. Re-drawn from Thrower (1962) Fig. 4.

import some label, as is revealed when large doses of $^{14}CO_2$ are fed to a source leaf, by faint images over mature leaves (Thaine, Ovenden and Turner, 1959). Presumably the effect is present with small doses but goes undetected. Thrower (1962) has shown that this faint image cannot be wholly accounted for by the fixation of $^{14}CO_2$ diffusing in the gas phase outside the plant for it persists even when mature leaves are sealed in chambers. In wheat (Quinlan and Sagar, 1962 and Lupton, personal communication) the transport into mature leaves seems to be relatively greater, though still slight. In tomato (Khan and Sagar 1966, 1969*b*) the import into mature leaves is substantial, especially at the beginning of fruit formation, and the ^{14}C imported is much more mobile and easily re-exported than that in a leaf which has assimilated $^{14}CO_2$. An explanation of this small reverse movement is offered below. Moreover, provided the sugar content of the vascular system is raised high enough by destroying the phloem locally, it is just possible to get enough sugar to move back into a darkened leaf to detect it by starch formation along veins (Mason, Maskell and Phillis, 1936*a*; Palmquist, 1938).

In a young herbaceous plant there are two main sinks, the shoot apex and the root system, of which the latter is much the larger, having a greater mass of respiring and synthesising tissue and no autotrophic capacity. The principal sources in such a plant, the mature leaves, are supplying both these sinks, and the traffic from each leaf to either sink can be assessed by following the distribution of labelled photosynthate from the leaf to both sinks. When this was done for *Soya* (Thaine *et al.*, 1959; Thrower, 1962) a rule emerged which we may term the third generalisation about patterns (*Rule 3*): that though label from all leaves goes to the root sink in greater amounts than to the apical sink, the contribution from the upper leaves to the apical sink is greater than that of the lower leaves, and the contribution of the lower leaves to the root sink is greater than that of the upper leaves. This raises a point of some interest and importance.

The implication that a translocation stream from upper leaves directed towards the root is crossing in the middle internodes a stream from the lower leaves towards the apex, does not require independent, opposing gradients of sugar in separate vascular strands, nor is it at variance with the view of a general declining gradient of sugar from source to sink. To consider the tracer movement and the sugar gradient contradictory is to misunderstand the relation of labelled to unlabelled sugar in a diffusion-type system, or to mistrust the completeness of the analogy between the translocation and diffusion systems. The concept of the translocation system developed by Mason and Maskell was that its mathematical description differed from that of a diffusion system only in having a higher value of the coefficient in equation 3.1. Therefore in predicting how label will behave in the system, we may replace the plant by a formalised diffusion system, and the only difference will be the speed of the changes. Consider then a tube of sugar solution with sinks of sugar at the top and bottom, that at the bottom removing sugar more actively. Replace the leaves by a number of sources of sugar in the form of reservoirs maintained at the same fixed concentration at several points along the tube. A steady state will be established with the 'leaf' region of the tube having a uniform sugar concentration and gradients declining from the uppermost leaf towards the apical sink, and more steeply from the lowest leaf towards the root sink (Fig. 5.2). Now, replace some sugar in the lowest leaf by labelled sugar (the total concentration being unchanged) and label will spread in the tube *in both directions* since diffusion is the summation of random molecular movements. The specific activity of sugar in the diffusion space will tend to uniformity; the channel within the boundaries of loading and consumption will eventually contain label at concentrations parallel to those of unlabelled sugar. The sinks outside the channel will accumulate label steadily; more will be accumulated by the root sink than the apical sink because this is both closer to the source of label and more active. Similarly if the label were in the uppermost leaf both sinks would accumulate label, but more would arrive at the apical sink than before since the distance of diffusion is less. Labelled molecules move under their own gradients in establishing an equilibrium irrespective of the unlabelled gradient in its steady state, or of the presence of a mass transfer of unlabelled material. This fact is of the utmost importance to the understanding of translocation phenomena, and, since misconceptions about the behaviour of tracer seem almost universal in the physiological literature, cannot be over-emphasised.

The other part of the system where crossing streams appear and tracer can move in either direction, an expanding leaf that can both import and export tracer, is susceptible of a similar explanation, and does not require separate phloem channels to be specialised for transport in either direction. The reservoirs on the postulated tube will equilibrate diffusively with the label in the tube, and that part of their space which is within the boundary into which sugar is loaded to maintain the concentration steady will eventually have the same radioactive concentration as the solution in the tube. If the loading space outside the diffusion space in a young 'leaf' contains sinks as well as sources, label will pass out of the diffusion space and accumulate in the loading space – label will pass into as well as out of the leaf. An 'older leaf' having no sinks will not accumulate label, but will contain a small amount – that in its diffusion space. If the reader thinks that these explanations in terms of the strict diffusion

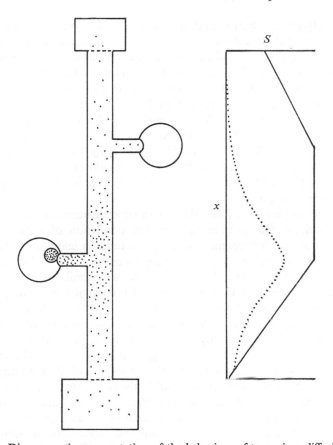

Fig. 5.2. Diagrammatic representation of the behaviour of tracer in a diffusion-analogue system. Sources (leaves) are represented by circles, and sinks by rectangles. The two sources maintain the same sugar concentration at their loading boundaries. The apical sink is smaller and less active than the root sink in that it maintains a higher steady sugar concentration at its boundary (shown dashed). A graph on the right shows the concentration of chemical sugar as a continuous line, and chemical sugar is in a steady state, i.e. its concentration is not changing with time. Tracer has been introduced into a part of the lower source and is still equilibrating by diffusive spread. Its concentration is represented in the diagram by dot frequency and on the graph by a dotted line. It moves independently of the gradient of chemical sugar in so far as the sugar diffusion coefficient is independent of concentration.

model are more far-fetched than the alternative polarised sieve-tube systems with different sieve tubes specialised for transport in different directions, let him reserve his judgement until the whole case has been presented, and register at this point the possibility of the present explanation.

A closely-related system is that of a major sink on the shoot system in the form of a developing fruit or infructescence, and for economic reasons the supply to the wheat ear has been much studied (Quinlan and Sagar, 1962; Carr and Wardlaw, 1965). Here the generalisation, our fourth is triumphantly simple: *Rule 4*, the fruit derives its substance almost entirely from the subtending leaf or bracts. Nearly all the assimilate from the flag leaf goes to the ear, and other leaves contribute almost none. In a precisely similar way the cotton boll is formed by the assimilation of its bracts (Brown, 1968). The developing lemon fruit draws its substance entirely from the leaves on that segment of the long shoot that bears it, and not at all from leaves on adjacent lengths of the shoot (Kriedemann, 1970).

The tomato plant is as unorthodox in its translocation patterns as in its morphology, but as even that strange confusion of leaves without axillary buds and infloresences without subtending leaves may be interpreted by a distortion of the ordinary rules of plant architecture, so with a little imagination the distribution of assimilates from tomato leaves can be seen as a twisted version of the usual rules. The patterns of assimilate movement from single leaves at different heights on the plant, and the changing patterns with age and ripening of successive trusses, have been thoroughly investigated by Khan and Sagar (1966, 1967, 1969a&b). They fed $^{14}CO_2$ to single leaves and assessed the amounts moved into roots, apex, stem, mature leaves and developing fruits, mostly after 24 hours, but also (1969b) in the first hours and after several weeks. In the young vegetative plant (1966) there was an apparent reversal of the usual Rule (3) in that the leaves near the apex exported a greater proportion of label down than up and the roots derived more from leaves 3, 4 and 5 than the nearer leaves 1 and 2. The stem was a strong sink, accounting for a large part of the export of all leaves; the apex, rather a weak one, and contributed to by all leaves. The leaves themselves retained a large part of the assimilated label, and increased in dry weight.

But Rule 4 was obeyed. As the successive fruit trusses developed they became the dominant sinks and a major proportion of the assimilated label went to them. As the plant grew taller and more complex each truss became the synthetic responsibility of the three or so leaves nearest to it. Each truss went through a period when it was a specially important sink and drew a large proportion of label from a large number of leaves, then its activity declined and the next took over.

Some part of the strange pattern of behaviour of the tomato may be related to its possession of what appear to be two separate transport systems, one specialised for upward, the other for downward transport. Like many Solanaceae it has phloem inside the xylem as well as outside, and Bonnemain (1965) shows radiographs suggesting that ^{14}C assimilate moves apically from a leaf only in the internal phloem, and basally (after 5 cm) only in the external phloem. This behaviour was not shown by *Datura* and, as will be seen in the next chapter, is quite untypical of most phloem which transports in both directions at once.

So far only intact plants have been considered, but early experiments on patterns of supply had to rely wholly on operations interrupting different paths of supply, and by way of helping to include such operations the tracer studies have been extended to see what happens to the normal pattern when the paths from some sources are blocked. These experiments usually consist of the removal of selected leaves. Thus in the plain her-baceous shoot–root axis, removal of all leaves except the source of label increases the amount of label moving to the apex, presumably because it is now no longer diluted by unlabelled assimilate from the other leaves (Thaine *et al.*, 1959). Further, removal of the basal leaves increases the contribution of the upper leaves to the root, and reduces the contribution to the apex (Thrower, 1962). Thus there is a kind of homeostatic switching mechanism which redistributes assimilates from the remaining sources to compensate for the lost ones. There is no need to envisage that switch as anything elaborate: the diffusion-analogue system will behave in just this way, establishing a new steady state with new gradients of unlabelled sugar, and label will be accumulated by sinks in the altered system at a faster rate since the specific activity will be higher. The fifth generalisation emerges (*Rule 5*)· that removal of sources diverts the burden of supplying the sinks on to the remaining sources, altering the patterns of movement. We can see at once why the early experiments not using isotopes were performed in vain. Now this generalisation has an important theoretical consequence, namely that there cannot be special channels connecting particular sources with particular sinks and specialised for transport in one direction; if immediate switching can occur all sinks must have channels of access to all sources, very much as in the diffusion model.

Another, and particularly elegant, example of the operation of this switch of sources is found in the work of Peel and Weatherley (1962) on the effect of defoliation on phloem exudates. Sugar from a single sieve tube was collected from the honeydew of an aphid feeding on *Salix* stem below the crown of leaves. The leaves had assimilated labelled carbon, and the honeydew contained labelled sugar at a steady level. Now on removing the leaves from the stem the sugar exudation remained constant, but the proportion of label in the sugar fell rapidly, indicating that the supply had switched to some other source which was unlabelled, and the obvious one would be the reserve carbohydrates of the woody stem.

The reader who has followed the argument on patterns of movement so far, and has considered the five generalisations with care, will already have performed the induction that leads to a single embracing rule that comprehends all the other five and answers the question with which we began. *Sinks are supplied by the nearest sources.* Or, expressed in another way, mass transfer is greatest over short distances. The rule is not intended to be strictly exclusive, in a sense each sink is supplied by every available source, but the contribution of the further ones is so strongly attenuated by distance as to disappear in comparison with that of the nearer sources. The first generalisation on longitudinal confinement cautions us that we may have to interpret our 'nearest' source a little liberally. A source that is close in distance to a sink may yet be relatively inaccessible if it is off the direct file of sieve elements leading to the sink. The second, on mature leaves and the change from import to export during leaf development, follows the rule faithfully. A very young leaf contains only sinks and will be supplied by the nearest sources, the youngest exporting leaves. As it begins to expand and develop photosynthetic capacity sources arise within the leaf which, by the rule, will supply the nearest sinks – those within the leaf. The leaf will not export assimilate. While the requirements of the leaf's sinks are greater than the output of the leaf's sources there will still be import from nearby leaves, but as soon as the leaf reaches such a size that its declining sink requirements are more than met by its rising source outputs, it will begin to export assimilate to the nearest sinks, the younger leaves. The overlap of import and export in Fig. 5.1 means that the supply from the nearest exporting leaves is not at first so completely attenuated by distance as to disappear in comparison with the internal supply, while the internal supply is still small.

The third generalisation of supply of the root and apical sinks by the nearest leaf sources is clearly in full accord with the rule; and the supply of some assimilate to the roots by the uppermost leaves, a consequence of the great activity of the root sink in comparison with the apex. Similarly the supply of a developing fruit by the nearest leaf is seen to be a special case of the rule. Indeed without having the experimental data we may predict with some confidence that this kind of relation will probably hold for many developing fruits; that the apple will be formed from the products of leaves on its dwarf shoot, coconuts from the spathe, and that persistent green calyces (as in the hazel nut) may contribute largely to the matter of the enclosed fruit.

The fifth and final generalisation on switching of sources when one is removed is another special case. If a leaf is supplying a sink and is removed, the nearest remaining leaf will take over the supply, and, if that leaf was already contributing to some second sink, its contribution to that will be reduced. The Peel and Weatherley switch in the supply to the aphid from leaf to stem reserve is instructive in showing an example of the modification of mere distance by the lack of direct longitudinal connection. The stem sources that are drawn upon when the leaves are removed are presumably close to the aphid, closer in distance than the leaves, but not on the direct file of sieve elements in which the aphid is feeding. Mass transfer from these sources while the leaves were present was therefore slight, but in the absence of the leaves, increases down a steepened gradient.

Armed with this rule we should be able to approach any new experimental arrangement of source and sink confident of being able to predict fairly closely the patterns that will be found. It would be pleasant to be able to turn at this stage to transport of some other matter than assimilate and show how patterns of, say, nitrogen movement follow the rule, but data are wholly wanting. Indeed none of the movements of substance other than carbohydrate has been studied in sufficient detail to stand as a test of the rule. The only general principle that emerged in Chapter 4 was that many of these other substances moved in the same general pattern as the carbohydrate, that is towards the sinks of carbohydrate. Nothing more exciting, then, can be deduced than that sinks of carbohydrate tend also to be sinks of weed-killers, minerals, viruses and amino acids, a lame and obvious conclusion.

Except for considering the conditions in which streams of labelled translocate appear to cross, a whole controversial section of the pattern story has been left out of this chapter and reserved for the next. This is the question of bi-directional movement; of whether a particular part of the translocation channel can carry two substances simultaneously in opposite directions. It has been omitted because this chapter is based on facts. The bi-directional movement controversy is a matter of speculation and polemic.

6. Bi-directional movement – the rival mechanisms

That the *sap* hath a Double, and so a *Circular* Motion, in the Root; is probable, from the proper Motion of the Root, and from its Office. From its Motion, which is Descent: for which the *sap* must likewise, some where have such a Motion proper to it. From its Office, which is To feed the *Trunk*: for which the *sap* must also, in some *Part* or other, have a more especial Motion of Ascent.

Grew (1682)

With the appearance of the 1928 volume of the *Annals of Botany* the argument about the channel that carried this large traffic of organic matter was stilled. The possibility that the movement might be through a wide area and at some easily-explained pace vanished, leaving physiologists with the task of explaining by what mechanism the sieve tubes could move sucrose at the rates listed in Table 1.1. In a sense they were only thrown back to the state of mind of those who followed de Vries (1885) in picturing the traffic as confined to the sieve tubes and effected by the circulatory streaming of protoplasm within each sieve element and diffusion through the pores of the sieve plates, a view that Curtis had maintained throughout the controversy of the 1920s. Now, however, that the magnitude of the problem was plain, there was an uneasy feeling abroad that the rates calculated by Dixon and measured in cotton could not be accounted for in this way. The cotton work revealed characteristics of the process which were of the right kind to be consistent with the de Vries streaming hypothesis in that the streaming motion provided just that 'acceleration of diffusion' which fitted so closely the experimental data (Chapter 3). Substances would move under their own concentration gradients at rates proportional to the gradients, accelerated by an amount depending on the speed of circulatory streaming. Thus far the model was of exactly the required kind. But the objections were two, and so serious as to blind many to the partial merits of the model. First there seemed to be no streaming motions visible in the protoplasm of mature sieve tubes. Many had looked and seen no streaming. This alone would not perhaps have been an absolute objection, for the fragility of the system was renowned and the failures to observe streaming could be blamed on the difficulty of keeping the sieve tubes functioning in sections thin enough to observe them. But the second objection appeared to rule the hypothesis out completely: that even were the streaming happening, protoplasmic movements at velocities known in other systems would not provide anything like the necessary acceleration.

Recent microscopic observations at this station on the phloem of cotton have shown, in agreement with the observations of Curtis (1929) and others on other plants, that protoplasmic streaming (up to 3 cm per hour) regularly occurs in the young sieve tubes near the cambium, and also, throughout the phloem in the elongated phloem parenchyma cells which run parallel with the sieve-tubes and, in cotton, comprise nearly half the 'sieve-tube groups'. We have not, however, observed any streaming in mature sieve-tubes with open pores. If streaming is restricted to phloem parenchyma cells and the young sieve-tubes it would appear, from the cell dimensions, that the maximum possible acceleration of diffusive movement, assuming diffusion across the cell-walls at the same rate as in water, is of the order of 1,000. We should require, however (cf. Mason and Maskell, 1928a & b), an acceleration of more than 20,000. Consequently, until we can obtain evidence of streaming within mature sieve tubes and also in and out of the sieve-pores, it seems that the question of mechanism must be left open.

(Mason and Maskell, 1934.)

This latter modification of the hypothesis, suggesting that transfer between sieve elements was accounted for not by diffusion but by an extension of the protoplasmic motions giving a two-way stream through a sieve plate, was a vain attempt to overcome the second objection on rates. It was not sufficient quantitatively, nor was it possible energetically (see Chapter 12). Discouraged by these formidable arguments, most workers on translocation neglected the direct and careful observation of functioning sieve tubes for thirty years.

There was already a rival hypothesis in the field. This had been proposed very briefly in 1926 by Münch and was elaborated in the following years to a book-length statement (Münch, 1930). Transfer, he said, was happening by the bodily flow of solution through the sieve tubes under pressure gradients. The pressure difference was provided by the turgor pressure of the sieve elements, higher at sources of sugar where reserves were mobilised and converted to smaller, osmotically-active molecules, lower at sinks where sugar molecules were removed from the sieve tubes to be respired or condensed. It was not necessary that the whole turgor of a source organ should be greater than that of a sink; it was sufficient that the turgor of the sieve tubes of the source should be greater. This could be achieved by active loading against a concentration gradient. The sieve pores were necessarily regarded as open – mere holes through which the flow passed – while the flow took place in the vacuoles. The protoplasm of the sieve elements served no purpose but as a lining to the sieve tubes confining the flowing solution. Water, which accompanies the sugar to the sink, must be either eliminated or returned, and Münch suggested that it was transferred to the xylem and flowed back from the sinks as part of the transpiration current. He even describes an experiment which, he claims, demonstrates this water. Flaps of bark separated from the wood were said to exude drops of water on the cambial surface; but other workers have failed to find any such excretion. This mass flow is just such a motion as was pictured by Dixon when he calculated the rate of movement of a 10 per cent sugar solution into a growing potato tuber, and we know (Chapter 1) that if it occurs its speed must be 20–100 cm hr^{-1}. Now this kind of mechanism will have properties fundamentally different from the streaming mechanism and will not show those close analogies with diffusion which were the theme of the Mason and Maskell results. For while the measured gradients of sugar concentration are in the right direction to drive the flow by turgor pressure differences, and while the rate of this flow will be roughly proportional to the steepness of the gradient, every solute will move in the direction of the pressure gradient, and not, as in the accelerated diffusion type, in response to its own concentration gradient.

Objections to the Münch hypothesis were of two kinds: those which denied the efficacy of the driving force, and those which denied the whole possibility of a bodily flow of solution. The calculations of resistance to flow contain many errors. The tangled evidence is perhaps best illustrated and resolved by quoting a passage from Mason, Maskell and Phillis (1936a):

The difficulty, however, is the *resistance offered by the sieve-plates to mass flow*. Crafts (1931, 1932) has in fact rejected flow through the sieve-plates on this ground and makes the fantastic suggestion that the walls will offer less resistance to flow. The fallacy in his calculations of wall resistance has already been exposed by Steward and Priestley (1932). Also the rates he observes for entry of dye solutions under pressure into bark (0.1 to 0.3 cm per minute for a gradient of 1 atmosphere over only 3 cm) show that for normal flow pressure gradients far higher than those he has rejected for the sieve-plates would be required. He appears to be equally wrong in his conclusion that the *dimensions* of the sieve-pores exclude mass flow during exudation. He *calculates* that the initial exudation rate of 4.05 cm per minute from the sieve-tubes of *Cucurbita Pepo* would require a pressure gradient of 20 atmospheres per metre and concludes that a pressure gradient of this magnitude cannot be accepted since the maximum osmotic pressure of the sieve-tube sap does not reach 20 atmospheres. But at the moment of cutting, and for a short time after, the pressure *gradient* must be many atmospheres *per centimetre*, i.e. greatly in excess of what is required. For normal transport in the intact plant the mass flow theory requires rates through the sieve-tubes of 0.3 to 2 cm per minute (Münch, 1930). These rates, on the basis of Crafts's measurements for cucurbit sieve-plates would require pressure gradients of 1.5 to 10 atmospheres per metre, i.e. gradients that are of the same order as those actually observed in trees and in cotton. It would appear that the dimensions of the sieve-pores, except perhaps in the fine leaf veins of Dicotyledons (Fischer, 1885) and in Conifers (Hill, 1901; Strasburger, 1891) are not in conflict with the mass flow theory *provided the pores are normally open.*

But almost all observers agree that the protoplasm of each sieve element, though it may be dilute and tenuous, covers the sieve plate and fills the lumen and pores (Schmidt, 1917; Crafts, 1931*b*; Schumacher, 1933; Kollmann, 1960; Esau and Cheadle, 1959, 1961; and Chapter 10). The reverse is stated by Duloy, Mercer and Rathgeber (1961) on the basis of a study with the electron microscope, but the weight of evidence seems against them. And if the pores are full of protoplasm very large pressures would be required to force solution through them. Indeed, a series of plates with narrow pores that are filled with a gel seems a system singularly ill-adapted to carrying a flow of solution, unless, as Spanner (1958) has suggested, the sieve plates are the seat of the driving force. Then the pressure gradient might be increased at each sieve plate, not diminished as in the Münch version of mass flow. This attractive hypothesis explains, indeed, how flow might be possible through the known structures, but the electro-osmotic forces which are substituted for the turgor gradient to drive the flow are not easily accessible to calculation or experiment. The mechanism requires only that the sieve plate should carry an electric charge, and that a potential difference should exist across each sieve plate. In order that the transport should continue down the file of sieve elements this potential must, of course, be polarised in the same direction for all consecutive sieve-tube elements of one sieve tube. Moreover it would have to be maintained at the expense of metabolic energy for the flow to persist, and for this double purpose, Spanner envisages an ionic pump transferring potassium ions between sieve-tube elements via the companion cells. Without much more precise information about the structures and detailed histochemistry of sieve elements, their protoplasm and pores, exact requirements and predictions to test this electro-osmotic hypothesis cannot be formulated, nor, in the present state of technical skill, can experiments be done on a fine enough scale (that is, on individual sieve plates) to test the predictions were they made.
    There are those, too, who go beyond mere dissatisfaction with the driving forces proposed for a bodily flow of solution, and claim that no mechanism of this class is satisfactory. That bodily flow of solution necessarily means that all solutes will move in the same direction and at the same speed, while they feel that there is sufficient evidence to show that solutes can be translocated independently of each other. It is these people who see the *experimentum crucis* in the demonstration that different substances can move in opposite directions simultaneously – the

bi-directional movement of our chapter title – and, being on the side of the controversy which is in the position of having a positive effect to demonstrate, have made several attempts. The mass-flow proponents, being in the more difficult position of having to show that bi-directional movement cannot take place, have refrained from this kind of experiment, contenting themselves with criticising the apparently positive results of their colleagues. It was the Trinidad cotton-research school who first attempted an experimental demonstration of bi-directional movement (Mason *et al.*, 1936*a* & *b*), feeling that their earlier work on carbohydrates, nitrogen and minerals had demanded more independence of movement than was allowed by the mass-flow hypothesis. They attempted to make nitrogen move upwards from storage positions at the base of the cotton stem at the same time as carbohydrate was moving down from illuminated upper leaves to the basal region whose leaves were darkened. The shoots of all experimental plants were isolated from the roots by removal of a ring of phloem above the cotyledonary node, and supplies of xylem nitrogen from the roots were excluded by water culture in solutions which finally contained no nitrogen. The object was to make the apical parts dependent for their nitrogen supply on the stores of the lower stem. In the Ringed group of plants a second ring separated the apical region with illuminated leaves from the basal region with darkened leaves; in the Normal group there was no such ring. Nitrogen was found to enter the apical region of the Normal, but not of the Ringed, group during the fourteen days of the experiment, indicating movement was by the phloem. Carbohydrate meanwhile passed in the Normal group from the apical to the basal region. The authors concluded, not unreasonably, that nitrogen and carbohydrate had travelled simultaneously in opposite directions through the phloem in the region of the upper ring.

Palmquist (1938) considered that there was a loophole in this demonstration and in the similar experiment of Fischer (1936–7), namely that 'their samples were taken at intervals of at least two days, and it is possible that the two solutes moved in opposite directions at alternate periods of time within the two-day period', and proceeded to plug this loophole with his own shorter-term experiment. This was a piece of work of some elegance employing a system that has considerable potential: the trifoliate leaf of *Phaseolus vulgaris*. This was chosen partly because the plant was one of very few whose tissues did not fluoresce with a colour that might be confused with his chosen second substance – Na-fluorescein (uranin) – and partly because the two lateral leaflets were found to be very closely congruent, with very similar dry weights. The one therefore acted as detached, the other as attached leaflet in a measurement of translocation of dry weight out of leaflets in the dark like those discussed in Chapter 1. 'An overnight loss in carbohydrates from one of these leaflets can be demonstrated easily by removing one of the pair in the afternoon and the other the following morning. In a series of leaves so treated, the average dry weight of the leaflets removed in the morning was about eight per cent less than that of the leaflets removed in the preceding afternoon.' Fluorescein was introduced into the terminal leaflet from a 1/1000 solution applied to the scraped distal end of the midrib, and spread in the phloem provided the plant was fully turgid. It moved, moreover, into the lateral leaflets at all times during the night while carbohydrate was leaving them. In a second, confirmatory experiment Palmquist induced translocation into the darkened terminal leaflet (as evidenced by starch formation around the veins) by scalding the petiole below the illuminated lateral leaflets and thus building up the sugar concentration of the petiole enough to reverse the gradient into the terminal leaflet. Simultaneously, fluorescein applied to the tip of the terminal leaflet moved out towards the scalded ring.

This did away with the alternate-streams-at-different-times objection, but left two which the mass-flow school were not slow to point out: that fluorescein was not a natural translocate; and that different sieve tubes were polarised for transport in opposite directions, that although bi-directional movement might occur in the same vascular bundle (there was only one in the *Phaseolus* petiolule) it did not happen in a single sieve tube. Such movement in a single sieve tube was much more difficult to establish and is still argued about today, even with the added power of tracer methods and apparently clear demonstration. These methods release us from the awkward substance-pairs, carbohydrate–nitrogen and carbohydrate–fluorescein, whose complications bedevilled the two investigations described above, but impose their own limitations on substance. In practice the phloem-mobile tracers are limited to four: ^{14}C, ^{32}P, ^{3}H and ^{35}S, and it is the first pair that has been used in simultaneous applications. Chen (1951) and Biddulph and Cory (1960) have shown that when one of these is applied to the leaf at the top, and the other to the leaf at the bottom of an internode, each spreads into the common internode independently of the other. This is precisely what we would expect from what we learned about patterns in the last chapter, and is consistent with the two experiments just described on bi-directional movement. Bearing in mind, however, the pattern-rule about longitudinal confinement in phloem and the complexity of connections of leaf traces with the vascular cylinder, we shall not expect the tracers to be spreading in the internode in the same sieve tubes or vascular bundles. Biddulph and Cory achieved sufficient resolution in their autoradiographs to show just this difference of path in the internode, ^{32}P spreading from an upper leaf of *Phaseolus* downwards in the one set of phloem bundles, while ^{14}C spread upwards from the lower leaf in another set. In other experiments, the vascular connections being right, the tracers seemed to share the same bundles. At this level of resolution we are still very far from demonstrating the two movements in one sieve tube.

Such a demonstration seems to have been made by Trip and Gorham (1968a). They fed ^{3}H-glucose to a leaf of a squash plant by vein injection, checking by micro-autoradiography that the label moved out of the leaf in the phloem. Then (100 minutes later) they fed $^{14}CO_2$ by photosynthesis to a younger leaf an internode or two above. After a further 20 minutes, when the two distinguishable sugar streams had overlapped in the petiole of the younger leaf and parts of the stem, the plant was frozen and processed for autoradiography. The resolution of their radiographs is just sufficient to show label confined to single sieve tubes on longitudinal sections. Both ^{3}H and ^{14}C were shown to be components of this label by exposures with and without masking. So here is a single sieve tube carrying labelled sugars that have come from opposite ends of the transport channel and have reached this spot only a few minutes before. The simplest explanation is surely that one sieve tube carries sugars in both directions. A dedicated opposition however does not accept this explanation, and maintains that the two sugars may have got to this sieve tube by another path than along it; that one at least of them has travelled in an adjacent sieve tube and moved sideways to appear in this one. If this has happened, it would be contrary to general experience that lateral movement between sieve tubes is much slower than along them.

There is another set of evidence that supports the view that a single sieve tube carries material both ways, but since this relies on techniques using aphid exudates, it will be treated in Chapter 8.

An explanation has already been given of how such tracer movements as these would be expected if the translocation system is truly analogous with diffusion systems, and the tracer evidence does seem to favour a mechanism of this type rather than one of the Münch type. For the sake of completeness rather than from intrinsic interest or merit, it is necessary to record those other proposed mechanisms which would behave like the diffusion system, but which invoke an accelerating force different from the streaming of protoplasm. These are the 'surface-spreading' mechanisms. They would employ the asymmetrical molecular forces existing at a phase boundary to transport sugar rapidly in the plane of the interface. The most complete mechanism of this kind was suggested by van den Honert (1932) who constructed a model interface between ether and water along which the speed of spread of a potassium hydroxide/potassium oleate mixture could be measured by colour changes of an indicator in the water layer. But the concept was not new; it was implicit in the proposal of Mangham (1917) who pictured the surface-spreading going on over the surface of cell particles with diffusion between the particles. These mechanisms have never gained wide support partly because no suitable phase boundary is known to exist, and partly because the non-polar sucrose molecule is a most unlikely one to be acted on by surface forces of the suggested kind. Furthermore, the dry weight masses of Table 1.1 spread out in a molecular layer would occupy a very large surface indeed.

How then should the controversy over bi-directional movement be regarded? In so far as it can be distinguished from the controversy over mechanism, its occurrence is only a special case of the patterns discussed in the last chapter, and, regarding the pattern evidence as a whole, it is my opinion that bi-directional movement is more likely than not. Certainly universally-accepted evidence demonstrating its occurrence in one sieve tube is wanting, and may not be obtainable for a long time yet. The most promising line would seem to be an extension of the fluorescent-dye experiment of Bauer (Chapter 4) to embrace two dyes moving in opposite directions while a single sieve tube was watched. The difficulty of introducing the dyes into the same file of sieve elements, a consequence of limited sideways movement and specialised vascular connections, might be insuperable. Nevertheless it seems that we should be prepared for a demonstration of the movement. In so far as the movement is regarded as a decisive test of mechanism it must be admitted that it has proved very unsatisfactory. The nearer the opponents of mass-flow have come to demonstrating bi-directional movement, the more complicated have the explanations of the defenders become. The evil consequences of the emphasis given to this argument have been that much thought and effort that should have been devoted to the collection of fundamental information have been wasted in devising, performing and criticising 'critical' experiments. The romantic appeal of the true *experimentum crucis* should not blind us to the fact that very rarely can one be devised and performed. As a method of research it is of far less value than the patient collection of coherent information.

7. The effect of temperature

The classification of the constituents of a chaos, nothing less is here essayed.
Moby Dick

No detailed mention has yet been made of influences that external conditions may have on translocation and such studies are valuable in the light they may throw on the process. For example, change of temperature might be expected to affect more profoundly a metabolic mechanism of the de Vries–Curtis class than a physical one like that of Münch. The temperature coefficient of the physical type, where temperature would have its effect mainly through such things as changing viscosity, might be between one and two, while for a metabolic process it would be higher, say two to three or more. Studies of the effect of temperature on translocation form one of the most confused and unsatisfactory sections of the subject. This arises from the almost complete lack of any reasonable quantity to measure translocation in, which lack in turn stems from ignorance of the process. Having no proper measure, it was impossible to find the effect of temperature on the measure. The mass transfer figures of Chapter I are more consistent and contain less assumptions than any other measures used by experimenters like growth, 'speed' of the hypothetical solution of equation 1.1, or rate of dye movement, and yet they are not so simple as to be wholly satisfactory for the assessment of the effect of a factor like temperature on the movement of substance. Consider that movement into and out of the transport channel may be affected as well as movement within the channel; that the rate of loss of carbohydrate from a source or of its arrival at a sink will be influenced by temperature both through the effect on transport and through the effect on respiration, higher temperatures of, say, the sink increasing respiratory loss to an extent that the apparent arrival of carbohydrate may be smaller than expected, or even absolutely smaller, at higher temperatures. Consider again that placing a whole plant at a different temperature may have so many inter-related effects on its metabolism as to make the disentangling of the temperature effect on one part of its physiology a hopeless task; that the local change of temperature on one part of a translocating organ is both difficult practically, so that many replicates of a treatment are awkward to arrange, and not certain to affect translocation only. Clearly the selection of a satisfactory measure will be a task of some delicacy. Further, the specific mass transfer figures contain, if we may generalise from the cotton data of Chapter 3, concealed quantities representing gradients of sugar concentration as well as the reactions of the plant systems to the different temperatures at which the measurements were made. For example, the value for specific mass transfer out of *Phaseolus* leaves recorded by Crafts (Table 1.1) during a day, is greater than the transfer out at night, and this may be because the temperature – which I have said in Chapter 1 was not specified – was higher, or it may be entirely an effect of a steeper sugar gradient out of actively assimilating leaves in the light, or both influences may be at work. Thus to separate these factors, to be as rigorous as even at this elementary stage of our knowledge seems desirable, measurement should be made of the influence of temperature, not merely on the specific mass transfer, but on the translocation coefficient, on K of equation 3.2. So we are two levels of sophistication beyond the measurements actually made. Moreover, measured as Mason and Maskell did it, K is subject to all the complications of respiration noticed above, since it is derived from a rate of loss or arrival of respiratory substrate combined with the gradient of respiratory substrate. The constancy of K in the narrow temperature range prevailing at the Cotton Research Station was a piece of great good fortune. To investigate variations in K by this technique over a range of temperature would be a massive experimental labour. If done, it could lead to a figure for the Q_{10} of K. Having said so much by way of preliminary, and having clearly in mind the kind of data we are looking for, we may turn to what measurements have been made to see if any useful information can be got from them.

The upper limit of temperature in which a translocation system functions has long been known as that danger-threshold for protein structure which destroys most vital processes, 40–50 °C. Indeed, the use of a steam jet to ring an organ without cutting it, to destroy phloem function without upsetting xylem function, and hence to distinguish phloem movement from xylem movement, has been a favourite experimental practice. Strasburger (1891) had applied steam jackets to stems in his studies of water movement in the xylem, showing that transpiration continued in spite of the heat, and soon after Czapek (1897) demonstrated the contrasting reaction of carbohydrate transport out through vine petioles, which the steam ring prevented. The steam ring is still used to verify that a particular movement is going on in the phloem not the xylem.* This clear, unquestioned property does not imply much about the system except that some kind of protein structure must be maintained uncoagu-

* A self-contained and portable apparatus for steam-killing small zones of tissue is described in Appendix 4.

lated if translocation is to be unimpaired. Both the 'accelerated diffusion' and the 'mass flow' adherents would claim this as consistent with their own views, for both picture some particular protoplasmic organisation to be necessary, the former for streaming or active surface, the latter for porosity to permit the flow, and as a pipe lining to confine it.

The low-temperature responses of translocation function are less well established, and introduce the confusion which has come from the want of a measure. Of these measures that of Child and Bellamy (1919) was the first and most primitive, which is not to imply that it is the least useful, for the field offers, as we shall presently see, archetypes of misguided sophistication. Primitive, in the sense that translocation was assessed without experimental labour by the development of dormant buds in the notches of *Bryophyllum* leaves and in the axils of *Phaseolus* and *Saxifraga* leaves. They showed that the inhibiting effect on the development of these buds by the main apex could be removed by jacketing the connecting organ with a cold (3 °C) ring – that isolation by cold had the same effect as physical severance, except that inhibition of the buds was re-established if the cold ring was removed in the early stages of bud growth. They interpreted their observations as meaning that the correlative factor, what we call a growth substance, 'depends for its passage from point to point upon metabolically active protoplasm, rather than upon purely physical transportation in the fluids flowing through "preformed channels" in the plant'. It is apparent that we are dealing here not with translocation in the sense of food transport, but with the transport of IAA or a related growth substance – a field of enquiry deliberately excluded from this book. The rule has been broken on this occasion for the reasons that this paper was often quoted in early studies on temperature; and that a phenomenon was recorded that will be constantly met with in the reactions of the system to low temperatures, namely the acclimatisation of the transport system to the cold block, the diminution of the effects of cold if continued for several days.

In most plants a temperature of 5 or 6°C., constitutes an effective block to the inhibiting action of the chief growing tip for several days, but these temperatures are near the upper limit of effectiveness for this species, and adjustment or acclimation of the cooled zone gradually occurs to such a degree that the buds which were at first physiologically isolated are again inhibited and cease to grow even before the low temperature is removed. When lower temperatures are used such acclimation may occur to some extent, but is less rapid.

Returning us to the theme of food movement, but still using growth as a measure of that movement, is a series of experiments employing darkened plants fed locally with sugar. The technique was first described by Weintraub and Brown (1950) and adapted by Swanson and Böhning (1951) as a measure of translocation. Seedlings of *Phaseolus* that had been kept in the dark for 1–2 days were fed through a primary leaf dipping in a solution of sugar, wetting agent and bactericide, and measurements made of the growth of the stem above the primary leaf node. The elongation of the stem after four or five days was proportional to the concentration of sucrose supplied, up to 0.75 M. Swanson and Böhning fed this highest sucrose concentration and placed a temperature jacket around the petiole of the fed leaf that allowed a range of petiole temperatures from 5 to 40 °C. Then the amount of growth was a measure of the effect of petiole temperature on translocation out of the lamina. The results of one such experiment are reproduced as Fig. 7.1. There is a strong increase of stem elongation with rising petiole temperature from 5 up to 20 °C, an optimum between 20 and 30 °C, and a falling off above 30 °C. The temperature coefficient between 10 and 20 °C is of the order of 1.5.

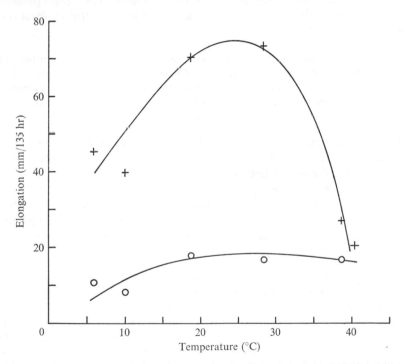

Fig. 7.1. Effect of petiole temperature on translocation out of the leaf as measured by the dark-elongation of the stem (in mm per 135 hours) when the leaf is fed with sugar solution: +——+. Control plants not fed with sugar: ○——○. Re-drawn from Swanson and Böhning (1951) Fig. 2.

These experimenters too, like Child and Bellamy, noticed the acclimatisation ('elongation rate increased with time at the low temperature level') and appreciated that with the long experimental time they may have underestimated the retarding effect. The observations were next extended (Böhning, Swanson and Linck, 1952) to show a similar effect on movement from primary leaf to the growing shoot of hypocotyl temperature, that is of a part of the plant not on the path of translocation.

The inference from the above data that phloem transport along a given path is influenced by temperature differentials applied outside that path is compatible with the view that translocation in the phloem is mainly by mass flow of solutions. Such a mechanism of translocation in the phloem would envisage a unit hydraulic system in which the effect of a given factor applied at any point would be transmitted throughout the entire system.

It is also compatible with any mechanism of translocation linked to the circulatory movement studied by Kursanov (Chapter 4) involving passage of assimilates to the roots and return to the shoot, and this explanation seems just as likely.

It is possible that elongation of the stems was limited by a compound synthesized in the roots. In this case, however, it would be necessary to further postulate that the synthesis of this compound was in turn limited by the rate of transport of a precursor from the leaf to the roots, for the rate of stem elongation in the minus-sugar controls was nearly negligible.

The dark-feeding, stem-growth method was then applied to tomatoes (Böhning, Kendall and Linck, 1953), where it worked in just the way it had in beans. Again the petiole temperature gave greatest elongation at about 24 °C; again the temperature coefficient in the range 12–24 °C was 1.5; but this time 'no physiological readjustment of the transport mechanism at low temperatures ... was noted'. Kendall (1952) enlarged the scope of the study to include bean-petiole temperatures that varied in daily cycles:

alternating low and medium petiole temperatures as used in the present studies were found in general to have a greater retarding effect on translocation than the constant low temperatures used by Swanson and Böhning, whereas alternating high and medium temperatures had a lesser retarding effect than constant high temperature. ... Evidently physiological re-adjustment to low temperatures was slower or less extensive under conditions of intermittent cold as used in these experiments. The greater amount of translocation which occurred when an alternating high and medium temperature was used as compared with a constant high temperature may be due to less injury to the protoplasm as a result of the shorter time of exposure to the very high temperature.

Measures of translocation by the rate at which exudates bleed from cut sieve tubes are probably unreliable because of damage to the cells, and we saw in Chapter 1 how the cut palm inflorescence gave a mass transfer in the exudate far outside the range of other measures. The method was used however by Crafts (1932) to assess the effect of temperature. Cucumber stems were cooled locally to 2–4 °C while the rest of the plant remained at ambient greenhouse temperature of 20–30 °C. Rates of exudation from the ends of stems cut near the cooling collar were about 40 per cent of the rates from normal stems, and the cooling collar was moved from one stem to the other so that each acted in turn as control and treated plant. Exudation from the region behind the cooling collar was normal. Reducing the temperature of the whole plant even by a smaller amount caused a greater reduction in flow, an extension of the finding that the reduction varied in proportion to the length of stem cooled. The composition of the exudate was unaffected.

Bleeding from the xylem under root pressure was the basis of another attempt at measurement in tomato plants. Went (1944) describes a 'bleedometer' for measuring the rate of xylem exudation; Went and Hull (1949) used it:

The rate of sugar transport is measured by recording how long it takes between the moment 7 per cent sucrose is applied to leaves and the moment that this sugar becomes effective in the root system. The activity of the root system is measured by the rate of exudation (bleeding) which requires respiratory energy... Since the rate of bleeding is variable even under completely controlled conditions, but varies parallel for all plants in one single experiment, the effect of applied sugar can only be measured by comparing the rate of bleeding of sugar-treated and control plants in individual experiments.

The reader will excuse me if I take him no further into this maze; there is nothing in there. Translocation may be difficult to measure but it is not that difficult. It is symptomatic of the tortuousness of these experiments that they record a result directly the reverse of all the others – that translocation is faster at low temperatures.

Bowling (1968) used the rate of accumulation of potassium by sunflower roots as a measure of the rate of arrival of carbohydrate from the leaves, and with a cold jacket (0 °C) on the stem between, showed a reduction of this measure of translocation. In some experiments there was evidence of recovery of the carbohydrate transfer while the cold jacket was still in place, in times between 5 and 24 hours; in others there was no recovery.

All the measures discussed so far have been indirect: translocation is assessed by some elaborate consequences of the movement. But more direct measures are possible and are of two kinds, those concerned with rate or speed and those concerned with amount or mass movement. Now of course these two points of view are related, but they are related through knowledge of the mechanism. One such relation is the mass-flow solution-speed equation 1.1, one that has been very commonly assumed to be useful but which has proved misleading. Another relation between the speed and the quantity components lies ahead; for the moment it is only necessary that they should be recognised when met with. In the preceding chapters the distinction is most clear between the spread of tracers in the investigation of patterns, and the mass transfer figures. As outlined in Chapter 5, tracer movement can occur at its own speed in a diffusion-analogue system whatever the mass transfer, even if there is none at all. So if we are to use both kinds of component as a measure of translocation we must be quite clear that they will not necessarily react in the same way to a change of conditions.

The measures that investigate the speed component do so by introducing a foreign substance and following its spread, by finding how far from the application it can be detected after a time and dividing distance by time. Since the foreign substance must be easily detected in small quantities, fluorescent dyes and radioactive substances have been commonly favoured. The theoretical and practical hazards of this measure will be discussed in a later chapter; at present the method will be taken at its face value because we are not particularly concerned with the absolute magnitude of the speed figure, only with the effects of temperature on it. Schumacher (1933, Table 1) records temperatures alongside maximum values of speed of fluorescein spread in different parts of *Pelargonium* and *Malva*, from which little consistent information can be gleaned beyond a general impression that in comparable parts higher temperatures favour higher rates. He devotes the next section to the influence of temperature, recording no speeds, but saying:

If one places samples of equal-aged, and as nearly as possible equally-developed specimens, the one in a cold room at 11°–12°C (bright cellar, cold thermostat and lighting), while the other is placed in a glasshouse or warm-chamber at a temperature of about 30°C, and applies the dye-gelatine on the leaf nerves at the same time, in the hot room after only about an hour the whole petiole has been passed, while in the cold, 3–4 hours may go by before the dye is easily detectable moving in the nerves.

We can gather no value of a temperature coefficient, only that the movement was 'ganz ausserordentlich temperaturempfindlich'. The effect was on transport rather than penetration to the transport system since it was manifested in plants held at the high temperature during application and penetration and then transferred to the lower temperature. The fluorescein measurement has been made more quantitative by experiments of Bowmer (1960) whose work gains added importance from its being on *Gossypium* and so providing data to supplement Mason and Maskell's results. Bowmer found that the height (internode length) of the cotton plant had a marked influence on the speed of fluorescein movement, so selected plants uniform in height. The temperature application was local, over part of the petiole of the source leaf and the stem, while the rest of the plant and the controls were kept at 25 °C. The results are reproduced as Fig. 7.2, whose general similarity to Fig. 7.1 is plain, with an optimum at 25 °C, a Q_{10} from 10 to 20 °C of about 2.5, and a strong falling off above 30 °C. The constancy of speed in the control plants inspires confidence.

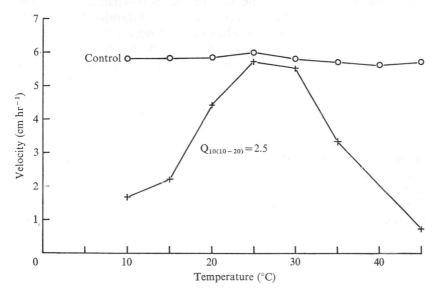

Fig. 7.2. Rate of movement of fluorescein (cm hr^{-1}) in *Gossypium* as a function of temperature of the petiole of the source leaf, $+$——$+$; and control plants at 25 °C, $\bigcirc$——$\bigcirc$. $V_{20°}/V_{10°} = 2.5$. Re-drawn from Bowmer (1969) Fig. 13.

Radioactive tracer for assessment of the temperature response of the speed component was first used by Vernon and Aronoff (1952) whose general method will be discussed below. ^{14}C was assimilated into first trifoliate leaves of *Soya* and translocation assessed by the speed of advance of detectable tracer along the stem. Temperature changes were achieved by surrounding part of the stem with a jacket filled with iced-water or water at 30 °C.

After 20-min. translocations, the radioactivity fronts (as determined by intercepts of typical radioactivity curves) for both chilled plants were 12 cm down the stem (8 cm below the 0 °C water level). For the check plant the activity front was 28 cm down the stem (24 cm below the 30 °C water level), showing that chilling the stems to 0 °C resulted in a drastic reduction in the rate of translocation.

One must refrain from using the data for deriving a temperature coefficient; the temperature range is very wide, the replication minimal and the method not unexceptionable.

Swanson and Whitney (1953) used tracers in a different way, not measuring the speed component, but assessing translocation by the amount of tracer that moved from application leaves. They dissected the treated plants into five parts and assayed each part separately for tracer, but finding no consistent picture, pooled the total translocated tracer to give a measure of translocation. Thus we must regard this as a quantity assessment, but an unsatisfactory one since sinks are not distinguished from path, stored tracer from mobile tracer. Radioisotopes of phosphorus, potassium, caesium or calcium were applied to a primary leaf of *Phaseolus*; temperature jackets maintained a length of stem or petiole at one of four temperatures from 5 to 45 °C. Great variability was found in the amount of tracer penetrating the leaf and entering the transport system and hence in the amount that arrived in the plant body, nevertheless the curves are of sufficient interest to deserve reproduction and a few are given in Fig. 7.3. An optimum is again apparent at about 30 °C, and the Q_{10} is often noticeably higher than in Bowmer's data between 10 and 20 °C. For the ^{32}P, ^{42}K and ^{137}Cs curves in Fig. 7.3 it is 1.85, 4.7 and 3.0. Calcium was practically immobile. In rather similar experiments with *Soya*, Barrier and Loomis (1957) record Q_{10}s for the movement of phosphorus and 2,4-D of 2.2 and 1.8.

Thrower (1965) could detect no spread of ^{14}C-labelled assimilate out of leaves of *Soya* in three hours when the whole plants were held at 2–3 °C. When only petioles of the source leaves were cooled, the assimilate moved in the same time up to the cooled zone and stopped there. When the cooled zone was warmed after two hours from feeding of label, the ^{14}C-assimilate now crossed the zone and spread to all parts of the plant in a further two hours. There was a lag of about 18 minutes, after which the labelled assimilate moved at about 17 cm hr^{-1}.

A measure that contains probably more of the mass transfer than the speed component is that of Webb and Gorham (1965) and Webb (1967), who measured the translocation of ^{14}C out of a squash leaf in 45 minutes following a pulse-feeding of $^{14}CO_2$ with first the node, and later various internodes held at a range of temperatures from 0 to 55 °C. This set of data, one of the most complete of its kind, is represented in Fig. 7.4 and is clearly similar to the previous three. There is a maximum at 25 °C, and the process is zero at 0 and 55 °C. Return to 25 °C from 0 °C allowed a slow (1 hr) resumption of translocation through the node, but the damage at 55 °C was irreversible. There was no sign of an adjustment and recovery at the low temperature in up to three hours. The later paper showed greater variability within and between plants in the central part of the temperature range, but confirmed that all parts behaved as shown in Fig. 7.4.

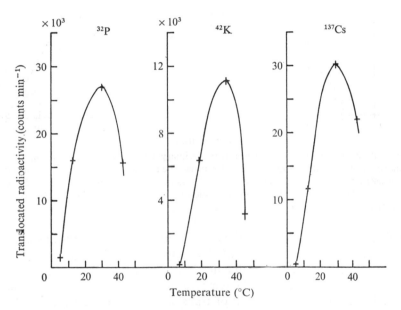

Fig. 7.3. Amount of tracer translocated in four hours out of source leaves of *Phaseolus* as a function of petiole temperature. Re-drawn from Swanson and Whitney (1953) Figs. 6, 7 and 8.

The pure quantity component, dry weight transfer, has been used comparatively little to measure the influence of temperature, but in the hands of Curtis and his colleagues provided our most consistent, care-fully-considered and revealing body of results. In Curtis and Herty (1936) loss of dry weight from bean leaves whose petioles were maintained at different temperatures was compared with the change (translocation assumed zero) in a leaf with a scalded petiole. Translocation was much reduced by cooling petioles to 2–4 °C, but as in experiments described earlier in the chapter, the effect diminished with time. Neglecting therefore

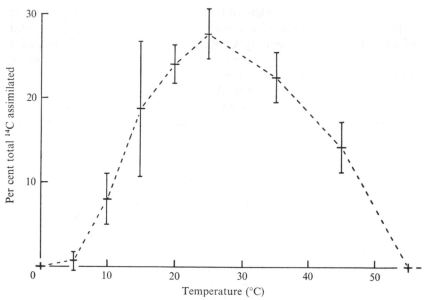

Fig. 7.4. Response of translocation from primary leaves of *Cucurbita melopepo* to changes of temperature applied to the lower petiole and node of the leaves. Translocation assessed as the percentage of total ^{14}C assimilated which moved out of the leaf in 45 minutes. The rest of the plant was held at 23–25 °C. Data of Webb and Gorham (1965) Fig. 1C, re-drawn.

their experiments lasting more than six hours, Fig. 7.5 records the data of their Table 4 where the loss of dry weight is graphed against petiole temperature. The regression line has been drawn and yields a Q_{10} of 2.0 over the interval 10–20 °C. The temperature did not rise to values where, on other evidence, a falling off of translocation might be expected. In a later paper (Hewitt and Curtis, 1948) the method is refined to take account of the effect of respiration which, as noted in the introductory remarks, will remove different proportions of translocate from leaf-sources and stem-sinks at different temperatures. Matched bean leaves gave values of initial dry weight and of final dry weight for attached and detached leaves held in the dark for 13 hours after a day of bright sunshine. Temperature regimes were not local this time; whole groups of plants were placed in rooms held at 4, 10, 20, 30 and 40 °C. The average of seven experiments, their Fig. 8, is re-drawn as Fig. 7.6 showing the separate influences of temperature on translocation and respiration. The optimum at 25 °C is now clear, and the Q_{10} from 10 to 20 °C has increased to 3 showing that the value in the earlier experiment when respiration was neglected is too low. Substantially similar results were obtained for *Asclepias syriaca* and

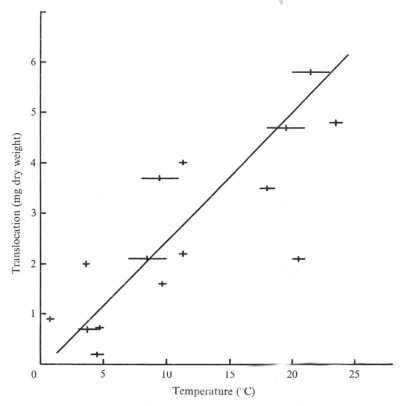

Fig. 7.5. Loss of dry weight in six hours from leaves of *Phaseolus* as a function of petiole temperature (mg dry wt less than leaf with scalded petiole). Data of Curtis and Herty (1936) Table 4.

tomato plants. Analyses of leaves for carbohydrates showed that the dry-weight losses could not all be accounted for in this form. The curves show how rapidly the respirational losses are going on at the higher temperatures, diminishing, as Hewitt and Curtis point out, the gradient and therefore the transport out of the leaves. They would account for the whole of the reduction in translocation at high temperatures by this effect, but we must be more wary. The experiments of Figs. 7.2 and 7.3 show just such optima when the transported substances are not respirable substrates and 7.4 shows one when the cooling is local. Either there must be some real reduction in the efficiency of transport at 25–30 °C, or the rate of transport of caesium, fluorescein, etc., must be linked closely to that of carbohydrate, an hypothesis for which we have met no other evidence.

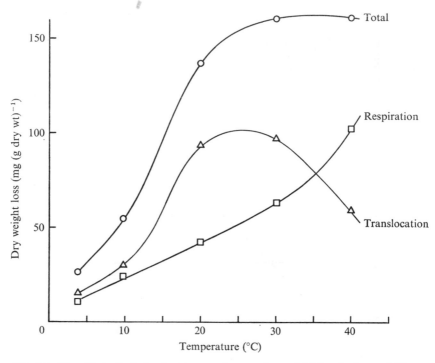

Fig. 7.6. The effect of whole-plant temperature on dry weight loss by leaves of bean plants during 13 hours' darkness in mg per g of initial or control dry weight. Total loss, ○——○; contribution from translocation, △——△; contribution from respiration, □——□. Average of seven experiments; re-drawn from Hewitt and Curtis (1948) Fig. 8.

In these experiments Hewitt and Curtis record no diminution of the low-temperature slowing of translocation with time, and a glance back at the experiments where the effect was recorded will show that the effect is characteristic of those where cooling is applied *locally* by means of a collar or jacket. It seems most likely that the effect is due to an increase of sugar concentration on the source side of the cooling collar giving a steepening gradient across the cooled region which produces an increasing rate of mass transfer. Swanson and Böhning (1951) are aware of the effect, though do not state its cause, and go so far as to impose the Hewitt and Curtis curve on their own diagram, remarking that the temperature response is much steeper in the range 5–20 °C because Hewitt and Curtis used a shorter time. If the present explanation is correct it would not have mattered what time Hewitt and Curtis used; since the whole plants were at uniform temperatures no duration would have shown any increase.

Swanson and Böhning's low-temperature reductions are smaller because they used cooling collars for the longer time; they would have achieved closer approximation to the Hewitt and Curtis curve by using a shorter time.

This wearing off of cold inhibition with time has been further investigated by Geiger, using that estimate of mass transfer noticed in Chapter 1 which depends on steady-state labelling of the translocate stream with ^{14}C, measuring the transfer of ^{14}C to a sink by continuous monitoring or destructive analyses, and calculating the dry-weight transfer using the specific activity. Local cooling of a 2-cm zone of petiole to 1 °C was used, and, with sugar beet, stopped translocation in a half-time of 4 to 15 minutes to the extent of up to 92 per cent. But with continued application of the cold zone, translocation resumed its former level again, with half-times of recovery between 30 and 100 minutes (Swanson and Geiger, 1967). The speed and completeness of this recovery seems to be a special property of sugar beet, since similar experiments with *Phaseolus* showed no recovery in nine hours unless the cold block was removed (Geiger, 1969a). In the same paper Geiger reports some observations by Bayer which show the two types of behaviour in ecotypes of a single species. A northern ecotype of *Cirsium arvense* from Montana 'recovered completely in less than three hours, after 8 cm of stem was held at 0.5 °C, while the southern ecotype from California showed no sign of recovery from a similar treatment of 8 cm of stem at 0.5 °C over a 3-hr interval.' No clear indication of what changes underlie either the inhibition or the recovery emerges from these studies. The idea proposed above of a steepened gradient across the cold block is rejected by Geiger (1969a) on the two grounds: (a) that there is no major over-shoot upon warming of the low-temperature-acclimated sugar-beet petiole (though there is such over-shoot when the sink-region cooling is released, Geiger 1969b) and (b) that the time course of recovery is the same whether the cold block is 2 cm or 8 cm long.

In an attempt to separate the effect on the speed component from the effect on the quantity component, Geiger and Sovonick (1970) came to the conclusion that in the sugar-beet temporary inhibition the effect on the speed component was the primary one.

The values for the temperature coefficient of dry weight transfer measured by Hewitt and Curtis are as close as the older literature affords to the ideal we started with, a Q_{10} of the translocation coefficient. When all those tendencies which co-operate to lower the measured Q_{10}, like the respiratory losses and the duration-reduction, are under control, the Q_{10} seems to be very large, 3 or more. How close this Q_{10} of dry weight transfer comes to our ideal Q_{10} for K depends on the constancy of the gradient of sugar out of the leaves. In carefully-matched plants in the same environment the gradients are likely to be closely similar and we may perhaps use Hewitt and Curtis's value as the Q_{10} of K without mistrust. The one direct estimate of the effect of temperature on K which will be discussed in Chapter 16, gives a Q_{10} of 2.8–2.9. The high value speaks loudly against a physical mechanism. Whether the speed component of the system differs in its response to temperature from the quantity component, the information is too slight to decide. Bowmer's data in Fig. 7.1 are the only reliable ones of this class and they show the optimum like the quantity-component data and a slightly smaller Q_{10} from 10 to 20 °C. The whole question of temperature response will be taken up again in Chapter 16.

8. Phloem exudates

The trees of the Lord also are full of sap.

Ps. 104

It is most remarkable, considering the amount of matter moving in the plants of Table 1.1 and the fact that these movements are going on all the time in all plants, that we cannot point at once to a sap, as we can point to an animal's blood, and say that there is the fluid carrying the solute. The plant saps familiar to the careful but untutored observer are the xylem sap that drips from leaves in guttation or wells up from a cut stem in spring, the latex that oozes from a plucked flower or leaf of some groups, and the resin that drips from damaged branches of others. These are obvious and abundant, but they have nothing to do with translocation. It is further remarkable that man, who by ingenuity and selection has made such use of translocated dry weight for food, harvesting grains, fruits, tubers and all kinds of sinks where dry weight accumulates, has made so little use of the material in transit. The examples of the use of translocate in motion are recondite indeed. Some peasants of Sicily, taught, it is conjectured, by the Arabs, harvest manna from *Fraxinus ornos* and *F. excelsior* by making cuts in the bark on summer mornings. Sap flows out and crystallises as it dries in the sun, forming icicles of sugary manna (mannitol) which contributes to the pharmacist's stock. The process is sufficiently unusual to have attracted not a little academic interest which is summarised by Hüber (1953) who watched the cutting and collection at Castelbuono. The collection of sugar sap from palm inflorescence stalks by the natives of Celebes provided the highest rate of mass transfer in Chapter 1 and is the only other example of direct harvest of translocate.

The excellence of the Palm wine or Toddy which is drawn from this tree makes however ample amends for the poorness of the fruit: this is got by cutting the buds which are to produce flowers soon after their appearance and tying under them a small basket made of the leaves of the same tree, into which the liquor drips and must be collected by people who climb the trees for that purpose every morning and evening. This is the common drink of every one upon the Island and a very pleasant one. It was so to us even at first only rather too sweet; its antescorbutick virtues as the fresh unfermented juice of a tree cannot be doubted. Notwi(th)standing that the Liquor is the Common drink of both Rich and poor, who in the morning and evening drink nothing else, a much larger quantity is drawn off daily than is sufficient for that use; of this they make a Syrop and a coarse sugar both which are far more agreable to the taste than they appear to the sight. The Liquor is calld in the Language of the Island *Dua* or *Duac*, the syrup and sugar by one and the same name, *Gula*. It is exactly the same as the Jagara Sugar on the Continent of India and prepard by only boiling down the liquor in earthenware pots till it is sufficiently thick. In appearance it exactly resembles Mollasses or Treacle only it is considerably thicker; in taste however it much excels it having instead of the abominable twang which treacle leaves in the mouth only a little burnt taste which was very agreable to our palates. The Sugar is of a reddish brown but more clear tasted than any Cane sugar I have tasted which was not refind, resembling mostly brown sugar candy. The syrup seemd to be very wholesome for tho many of our people eat enormous quantities of it it hurt nobody, only gently opning the body and not as we feared bringing on fluxes. (Sir Joseph Banks: Visit of the *Endeavour* to Savu in September 1770. Referring to *Borassus flabellifer*, Fan palm or Toddy tree.)

The tradition of the making of an extremely fibrous and indigestible bread from the inner bark of pines and elms by the primitive inhabitants of Europe in times of desperate famine survives – along with a few of their loaves – as a reminder that our ancestors were tougher than we (Nordhagen, 1954). But if man has been backward in exploiting this source of food, other animals have done justice to its abundance. Phloem-biting coccids, mites, aphids, etc., infesting summer trees may remove a large proportion of the sugars manufactured by the leaves. Not that they make use of the sugars; they use primarily the nitrogenous part of what they extract from the tree, excreting the excess sugar as drops of honeydew which fall as a fine sticky rain or may be collected by ants and bees. Büsgen (1891) showed that the rain of honeydew came from aphids and that the aphids penetrated the sieve tubes with their mouthparts. (See Esau, 1961.) So the sap is there, but it was not until 1860 that its relevance to translocation was realised by Hartig:

More important, in the past summer I have succeeded for the first time in demonstrating the primary building sap returning out of the leaves to the lower parts of the plant, and contained in the sieve-tube tissues of the bast sheath. Moreover it was in so simple a way that it is wonderful that the method did not become known a long time ago. Maples, oaks, beeches, hornbeams, limes, acacias, cherries from one to six inches stoutness, scraped in summer and autumn with the point of a knife in a horizontal or oblique direction as far in as the innermost bark layers, show directly a few drops of a watery sap which can be collected with a brush and obtained pure in sufficient amount to be investigated. It contains small quantities of nitrogen-containing substances that separate out on heating, no gum, but between 25 and 30 per cent sugar of many different kinds and crystal shapes, some sweet and having crystals like cane sugar, but also separating in exceptional crystalline shapes (e.g. in acacias in sphere-like tetrahedra), some tasteless (mannitol). Only the extracted sap of maples is in a high degree bitter and dries to brown drops like cherry-gum.

The flow lasted only a few minutes. Cuts at the top of a tree stopped exudate coming from subsequent cuts at lower levels on the same side, but cuts made successively from the bottom upwards yielded each time a fresh flow. He announces proudly: 'Without doubt we have here a transport sap going downwards. That it comes from the sieve tubes, direct microscopic examination shows.'

There is then a rich sugar solution associated with the sieve tubes and transport, and from observations of this sap, as Hartig saw, surely much can be learned about translocation. This chapter is concerned with the questions: What evidence can the sap provide about the translocation process? and How far can this evidence be trusted? In particular, is it a transport sap in the strict sense of a flowing solution such as is defined by equation 1.1? If not, what relation does it hold to what is moving? Questions which were strangely neglected from the time of Hartig's record until the 1930s.

Hartig's method of collecting the sap by cuts in the inner bark was used unchanged by Münch (1930), Pfeiffer (1937), Hüber, Schmidt and Jahnel (1937), Ziegler (1956) and Zimmermann (1957a). All mention the danger of losing the sap by cutting too deep. If the wound penetrates the wood the sap may be sucked in by the release of tensions there. The early studies in this list were concerned mainly with the osmotic pressure of the sap and how this varied with height of the tree, in attempts to find gradients of pressure that would drive the mass flow proposed by Münch, and for such studies a simple cryoscopic or refractometric measure of the solutes of the sap was sufficient. Nor were special precautions taken to preserve the sap unchanged, until interest in its chemical composition succeeded to interest in its osmotic pressure. Ziegler (1956) notes how quickly changes in composition of the sap occur, and stored samples in the cold and with preservatives. But, however carefully preserved the solution, there is still the suspicion that changes may have occurred, particularly in its water content and by eluting solutes from the bark, while the drops of sap accumulated on the tree. Moreover the precise origin of the sap is uncertain since the cut opens many cells. A very important and welcome advance in the technique of collecting the sap which minimises these errors was made by Kennedy and Mittler (1953). They found that if the phloem-feeding aphid with its stylet penetrating a sieve tube had its proboscis cut, the stump would often continue to exude sap for several days (Plate 4). Here was a notable refinement; the sap was coming from a single sieve-tube element, however it might arrive there; the sap was conveyed cleanly to the collecting pipette without the possibility of picking up adventitious substances on the way; and the flow continued for long enough to permit extended measurements. All was not pure gain, however, for quite few aphids yield satisfactory flows and they are restricted to few host plants. The two best are *Tuberolachnus salignus* on *Salix* and *Longistigma caryae* on *Tilia americana* (Zimmermann, 1961, but *Picea* and *Heracleum* (Ziegler and Mittler, 1959) and *Cupressobium* on juniper (Kollmann and Dörr, 1966) have also yielded aphid exudates. Microscopic study of the penetrating mouthparts has shown that the stylet always ends in a sieve tube when flow has been established (Plate 5) (Kollmann, 1965; Hennig, 1966; Evert, Eschrich, Medler and Alfieri, 1968). Anaesthetising of the aphid by CO_2 has been used to prevent the stylet being withdrawn when the cut is made, but in the author's experience is unnecessary. With *Tuberolachnus* selection of the right size of aphid contributes more to success. Very small aphids are docile but the flow through the small stylets is minute; very large aphids are active and often partially withdraw their stylets while the cut is being made; medium-sized aphids may usually be cut so as to yield a satisfactory flow. The maintenance of the flow is something to a large degree beyond the control of the experimenter, as is also the precise location of where the aphid inserts the stylet. A liberal supply of water and low saturation deficit contribute to the former; but the reasons why aphids will feed on one part of a stem and not another are best known to themselves. Measured flow of $1-2$ μl hr^{-1} out of the aphid-stylet food canal – a tube whose diameter was 0.6 μm at the tip and 1.8 μm at the proximal end – allowed Mittler (1957a) to calculate the pressure which must be driving the flow as 20–40 atmospheres. He was particularly concerned to show that the aphid was not sucking, but passively being fed; on the side of the plant the figure is of the first importance as a direct estimate of pressure inside the sieve tube.

Herbaceous plants do not usually yield any appreciable exudate when the phloem is cut, but many *Cucurbita* species do bleed a solution from cut shoots which has been much experimented on. Crafts (1939a), by letting the stem exude into water and observing the cut end, was able to see that the streams of sap, distinguishable by their density, came from the phloem not the xylem, and claims that all plants show this kind of exudation when actively assimilating. The large seaweed *Macrocystis* yielded abundant exudate which must have come from the sieve cells since the plant has no other obvious vascularisation.

Analysis shows the main solute of the exudate is sugar, almost always sucrose, but exceptionally the higher galactosides noted in Chapter 2, and in concentrations from 10 to 25 per cent. Reducing sugars are almost entirely absent when trustworthy techniques are used (Wanner, 1953*a*). If the exudate is intimately connected with translocation its other constituents may convey some message, and have been several times investigated. The various estimations are assembled in Tables 8.1 and 8.2. The exudate from *Cucurbita* is generally regarded as unrepresentative, both of exudates as a class, and probably also of the sieve-tube sap of *Cucurbita*. Its low sugar and high protein content have led to the suspicion that the sap is diluted with adventitious water as it bleeds, and also that a large part of the sieve-tube protoplasm is swept out with the sap through the large sieve pores. Calcium is supposed to be absent from exudates, and this absence is associated, as already discussed, with the immobility of calcium in general translocation. Mason and Maskell (1931) found no sign of its movement; radioactive calcium was immobile in the experiments of Swanson and Whitney (1953); and Ziegler (1956) goes so far as to point to the danger of the inactivation of the enzymes of sugar isomerisation by calcium ions and to the frequent occurrence of calcium trapped as oxalate in phloem. But quite plainly the exudates measured by Moose contained as much calcium as potassium, an element that Peel and Weatherley (1959) and Tammes (1958) regard as present in exudates in particular richness. Much of this variability may be due to seasonal changes. One constant feature of the sap seems to be its hydrogen ion concentration, slightly on the alkaline side of neutral. This is a pH which might inactivate enzymes like invertase, but Ziegler (1956) showed that whatever the pH of the sap no reducing sugar was produced while it was kept sterile, so the enzymes were absent not inactive. However, that this alkalinity is probably a seasonal rather than a constant character of the sap is shown by Eschrich (1961). The pH fell to 4.8 in sap obtained in minute amounts in January from *Tilia tomentosa*, and a similar fall was recorded for *Acer*. The exudates were alkaline again in summer, so that the appearance of alkalinity in Table 8.1 is a consequence of the experimental convenience of collecting sap in summer and autumn. The phosphate of the sap seems to bear no special relation to the sugar (Ziegler, 1956), and therefore is probably not formed by the breakdown of sugar phosphates which several authors state to be absent, but which have been recorded in *Cucurbita* exudate (Eschrich, 1963). Also we have seen that ^{32}P is fairly mobile between phloem and xylem.

The differences between the total and protein nitrogen columns of Table 8.1 represent the soluble nitrogen of the sap which is composed of a variety of amino acids. What information is available about the nature and amount of these compounds has been assembled in Table 8.2. Mittler (1953) records the presence of individual amino acids but says nothing about their relative amounts. In so far as the order in which he

TABLE 8.1 *Estimates of the composition of sieve-tube sap (amount of solute in mg per litre)*

Author	Plant	Sugar	Total nitrogen	Protein nitrogen	K	P	Ca	Mg	pH
Tammes (1958)	*Arenga saccharifera*	150,000	410	–	1200	100	10	96	–
Moose (1938)	*Fraxinus americana*	80,000	3000	–	800	–	800	300	7.1
	Robinia pseudo-acacia	150,000	500	–	900	–	900	300	7.2
	Platanus occidentalis	100,000	280	–	900	–	1000	250	7.3
	Cucurbita maxima	7000	–	–	200	–	100	80	8.0
Peel & Weatherley (1959)	*Salix viminalis*	40,000 to 250,000	15,000† to 25,000	–	2000	–	0	–	–
Cooil (1941)	*Cucurbita pepo*	5000	35,000	30,000	–	–	–	–	7.0
Ziegler (1956)	*Quercus rubra* L*	18,000	1400	900	–	60	–	–	8.0
	Quercus robur	1400	900	350	–	–	–	8.2	
	Robinia pseudo-acacia	14,000	2750	2000	–	720	–	–	8.4

– Not determined.

* Ziegler calls this *Q. borealis* Michx.

† This information is not in the paper, but has been kindly supplied by Dr Peel. The determinations were made in January and February.

TABLE 8.2. *Amino acids found in phloem exudates*

Amino acid	Mittler (1953)* and Peel & Weatherley (1959)† *Salix* Autumn and winter	Mittler (1953)* and Peel & Weatherley (1959)† *Salix* Summer	Eschrich (1963) *Cucurbita ficifolia*	Ziegler (1956) *Robinia* Autumn	Ziegler (1956) *Robinia* Summer
Aspartic acid	+	+	+++++	+++	+++
Glutamic acid	+	+	++	+++	++++
Serine	+	–	–	++	+
Threonine	+	–	–	–	–
Alanine	+++++	–	++	++++	+++
Valine	+	–	–	++++	++++
Leucine	+++	–	–	+++	++++
Phenylalanine	+++	–	–	–	–
Asparagine		++	+++	++	++
Glutamine	+	++	+	++	++
Citrulline	–	–	+++++	–	–
Proline	–	–	–	+++++	++
Lysine	–	–	–	++	++
Methionine	–	–	–	+	+

– Not recorded.
* Mittler recorded presence only, no estimate of amount, but the first ten acids are written in the order he gives them.
† Dr Peel has kindly supplied the list of acids he found in the exudates. It is identical with Mittler's list. The winter period was from October to April; the summer, the rest of the year.

wrote them down may reflect their relative abundance, this has been preserved in the table. Thus aspartic and glutamic acids perhaps rate more than a single + in Mittler's column by their having been named first. Peel and Weatherley (1959, and personal communication) recorded the same acids and amides and also made no estimates of amounts. The + signs in the other columns have been copied from the two papers named. There is no suggestion that the scales of comparison are the same beyond the fact that each gives 5+'s to the most abundant compound. Aspartic and glutamic acids are strongly represented in all saps investigated, and must be regarded as the major forms of soluble nitrogen, though there is clearly much seasonal and specific variation. Citrulline, high in *Cucurbita* exudate, was found fairly widely by Ziegler and Schnabel (1961), occurring in *Betula, Alnus, Carpinus, Ostrya* and *Juglans* exudates, and in concentrations up to 2000 mg l^{-1}.

There are in addition many substances whose occurrence has been occasionally recorded in exudate, and while, from the reports being diffused in different papers, one derives the impression that they are of sporadic occurrence, yet they may be fairly generally present but only intermittently tested for. Thus DNA was present in *Robinia* exudate at a concentration of about 5 mg l^{-1}, with rather less RNA (1.5 mg l^{-1}) and a considerable amount of adenine (62 mg l^{-1}) (Ziegler and Kluge, 1962). Allantoin (up to 1700 mg l^{-1}) was widely spread in *Acer* species together with some allantoic acid (Ziegler and Schnabel, 1961) which might have been, but probably were not, picked up from the bark tissues as the sap came out. Considerable amounts of growth substances were demonstrated in exudates from several trees by *Avena*-test assays (Hüber *et al.*, 1937). Enzymes of the sap are important in our deciding what it is and whence it comes. Phosphatase which will split ATP and fructose-1-6-diphosphate was shown by enzymic tests (Wanner, 1953*b*) and histochemically (Wanner, 1952), but the only other enzyme of the sugar–isomerism–glycolysis group was phosphoglucomutase. Notwithstanding the paucity of the enzymes, the sap showed a slight consumption of oxygen (20 μl O_2 g^{-1} fw hr^{-1}). No carbon dioxide was evolved, and any produced must have been retained in the alkaline solution. Organic acids, particularly citric and tartaric, balanced the high potassium content of aphid exudate from willow (Peel and Weatherley, 1959).

The water in the exudate demands attention, not for its presence or its amount, but for its origin and constancy. Many kinetic arguments about the behaviour of exudates and the inferences that can be drawn therefrom are based on changes in concentration. All such must be suspect if the water content of the exudate has changed from what it was when the solution was part of the living phloem. From what has been said of its collection and composition the source of the exudate must be the lumina of sieve tubes. Except perhaps in *Cucurbita* the absence of proteins speaks against the presence of much of the protoplasm of sieve tubes. The precision of the aphid-stylet extraction necessitates that that solution at least comes *via* a single sieve element and the first of the flow must be from the lumen of that element. The origin of the continued supply of exudate is less clear. By analogy the source of exudate from a cut is most probably also the sieve-tube lumina, but is its concentration unchanged? It is most probable that the concentration is less, the water content higher, in the exudate than in the vacuole. Everything points to the osmotic potential of the sieve-tube sap being more negative than the water potential of the nearby cells and a 20 per cent sugar solution let loose among the tissues of the phloem and cambium will almost certainly absorb water from many of the cells until it is diluted to equilibrium with their water potentials. Such dilution, revealed as the progressive weakening of the sap with time, was shown to occur by Tingley (1944) and Zimmermann (1957*b*) and investigated more fully by Zimmermann (1960, 1962). The ash among trees may be compared to *Cucurbita* among herbs in having wide sieve pores, and in yielding a flow of exudate which lasts an hour or more. When two incisions were made simultaneously two centimetres apart in the bark of *Fraxinus americana*, the exudate concentrations from both drifted downwards at the same rate (Fig. 8.1). The curve from the upper incision is above that from the lower because it is drawing on tissues above, higher up the sugar gradient, while the lower is drawing on sissue below the cuts. The dilution is a local phenomenon for if the tecond cut is made above and 20 minutes after the first, when sugar concentration has built up in this region of the phloem due to the effect* of the first cut (Zimmermann, 1960), the concentration of the exudate starts high, but drops quickly into place beside the lower one (Fig. 8.2). Thus 'while the concentration of exudate decreases *at* an incision, it increases at a distance of as little as two centimetres *above* the incision. This means that the dilution phenomenon is quite a local one. But it may also mean that water moves somewhat faster toward the incision than the solutes.' The protection afforded to the solution passing up the food canal of an aphid stylet by the walls of the tube is probably sufficient to prevent its dilution, since insect cuticle is so little permeable to water, and it is likely that we can trust the concentration of exudate collected thus, particularly as it shows no change with time (Fig. 8.3). A check on this concentration may be obtained from the pressure exerted to drive the flow through the aphid stylet. This is the turgor pressure of the sieve

*Reference should be made to the original paper if interpretation is desired of this rise, which is complicated and doubtful.

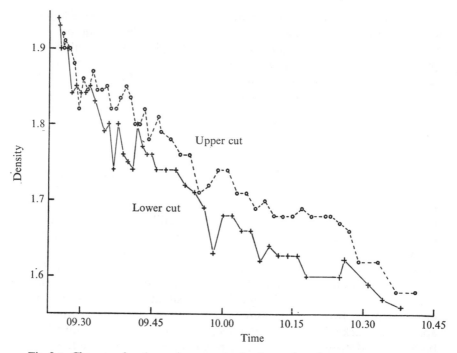

Fig. 8.1. Changes of molar sugar concentration in exudates from two cuts one above the other and 2 cm apart. Exudate from upper cut ----; from lower cut ———. Re-drawn from Zimmermann (1962) Fig. 2, using data supplied by the author.

element, and if the element can be made fully turgid, as it mostly must be to achieve exudate flow, the pressure will also be the osmotic pressure of the vacuole contents. The necessary pressure calculated by Mittler, we have seen, rose to 40 atmospheres, implying a 37 per cent sucrose solution. It must be remembered that a 10 per cent uncertainty in the radius of the food canal will produce twice this uncertainty in the pressure estimate, since the pressure depends on the fourth power of the radius.

Considering now the source of the exudate water, the rate of flow from the stylet, say 1 μl hr^{-1}, can be used to calculate what this implies about a rate of flow of solution longitudinally in the sieve tubes, if this is the way the solution is coming. Thus Weatherley, Peel and Hill (1959) regarding the sieve tube as a sealed pipe: 'Taking the diameter of the sieve tube as 23 μ, this represents a rate of movement at the site of stylet insertion of about 1 m/h if the sap is considered to converge on the stylets from both directions.' On the other hand it is not likely that the water is all arriving at the tapped element lengthways in the one sieve tube; Weather-

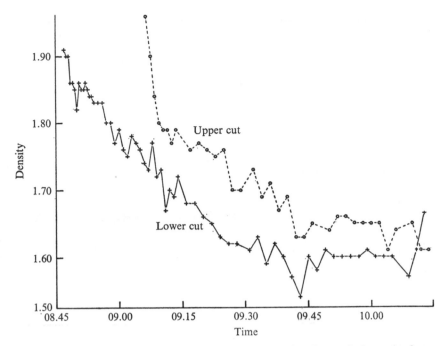

Fig. 8.2. Similar to Fig. 8.1. Second cut made 20 minutes after and above the first.

ley *et al.* go on to show that the osmotic pressure of the xylem sap affects the exudate from the stylet. Solutions of sucrose or mannitol of higher osmotic pressure than the xylem sap, forced into the xylem of a shoot, reduced the volume flow of exudate from a stylet in the bark, but at the same time the sugar concentration in the exudate increased, though not enough to keep the sugar transfer constant. It was as though the sugar was continuing to arrive lengthwise at the tapped sieve element, but that less water arrived radially up a smaller gradient of suction pressure between the sieve element and the xylem. This suggests that the water of the exudate comes from the xylem, not lengthwise with the sugar; that the two major components of the sap are of different origins and may vary independently. Similar and more rapid responses were demonstrated when the strip of bark was inset in the wall of a plastic tube and the xylem sap was simulated by the contents of the tube. Additional evidence for the independent movement of solute and water, and for the water of the exudate coming very locally to a cut from the xylem, has just been referred to in Figs. 8.1 and 8.2. Peel has further confirmed that the water in the exudate comes radially and not with the solutes (Peel, Field, Coulson and

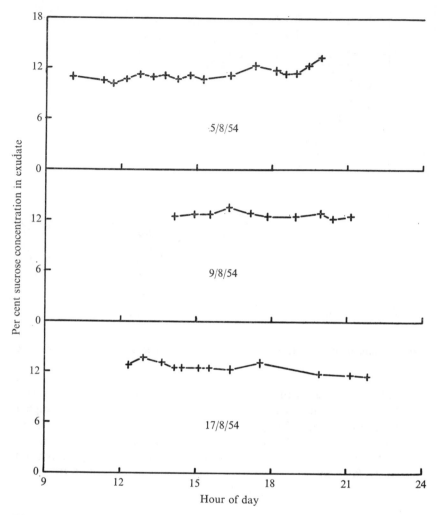

Fig. 8.3. Sucrose concentration of aphid-stylet exudate during a day following cutting. Refractometer measurements for three different plants, re-drawn from Weatherley, Peel and Hill (1959) Fig. 3.

Gardner, 1969; Peel, 1970). In the first of these a bark strip was placed over a divided water supply in two compartments A and B. Aphid stylets were established in the bark over each. ^{14}C-sucrose and labelled sodium phosphate supplied to the solution of compartment A moved along the bark and appeared in the exudate from the stylet above B, 2 to 6 cm away. However tritiated water did not. It flowed easily from the near stylet above A, but did not move along the bark any faster than by diffusion in dead strips. The second of the two papers demonstrates a similar difference in whole stem pieces fed with tracers at one end. Honeydew was collected from aphids feeding on the bark near the supplied tracers ($^{35}SO_4{}^{2-}$, ^{32}P-phosphate and ^3H_2O). A girdle in the phloem near the collection site, and on the side away from the tracer source, caused a rise in the specific activity of ^{35}S and ^{32}P as their dilution from sources in the rest of the bark was prevented, but the ^3H_2O remained the same, showing that it had not been so diluted. It seems that great caution should be exercised in drawing conclusions from concentration changes in the sap, since such may be rather changes in water content than in solute content.

How easily confusion can arise in kinetic studies is plainly seen in the investigation of the effect of illumination of willow shoots on exudate from aphid stylets in the bark (Peel and Weatherley, 1962). The sugar concentration fell during the dark periods, whether as long as twelve hours or as short as two; at the same time the rate of sap and sugar flow rose. Peel and Weatherley recall the effects of xylem osmotic pressure and go on:

This suggests that the effect of light and dark periods upon a potted cutting may be due to the accompanying rise and fall of the DPD of the xylem water. During the initial light period the stomata will be open, and transpiration will be proceeding rapidly. Thus tensions will arise in the xylem water. During the dark period, the stomata close, transpiration declines, and the tension falls. In support of this suggestion, it should be added that the water uptake due to transpiration by a leafy cutting which had been severed just above the roots was three times greater in the light than in the dark.

The suspicion was confirmed when fluctuations of xylem tension were prevented either by enclosing the leaves in a glass chamber, or by removing the root resistance – severing the shoot from the roots and placing the end in water. Under either of these conditions alternating light and dark did not affect the stylet exudation.

One reason for this project was that the authors hoped to demonstrate a rise in the sugar content of the sap as sugar produced in the leaves in the light moved down the stem, manifesting in the exudate the general rise of sugar content of the sieve tubes that Mason and Maskell had shown. They failed, as Mittler (1957*b*) had done, to show any such rise once the influence of xylem tension was removed, notwithstanding the several reports of such demonstrations being successful in tree-incision exudates. The first and most convincing of these is the experiment of Hüber *et al.* which is reproduced in Fig. 8.4. Exudates were collected from incisions in the bark of a tree of *Quercus rubra* at five stations down the trunk from

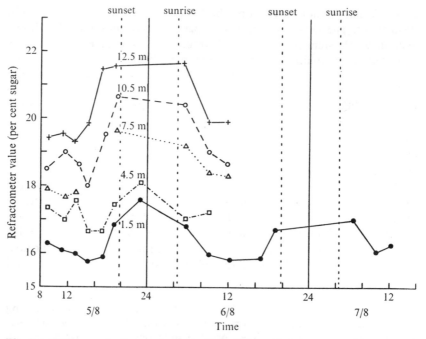

Fig. 8.4. Daily march of sugar concentration of sieve-tube sap measured as refractometer values at five heights on a red oak. Re-drawn from Fig. 4 of Hüber *et al.* (1937) with the addition of calculated times of sunrise and sunset at Tharandt for the dates of the experiment.

12.5 m to 1.5 m height over periods of several days and the sugar concentration assessed with a refractometer. The exudate concentration is high at night, declines in the early morning and rises again in the afternoon, and there is a suggestion in Fig. 8.4 that the changes happen first near the leaves and are detected successively at each station down the trunk. The concentrations change in the opposite direction to that expected if the cause were simple xylem-tension changes of the kind found in willow – if this were the mechanism the concentration of exudate should be higher in the day when the xylem tension is greatest. Hüber *et al.* consider carefully this and another possible origin of the changes, consumption of assimilates in passage, and conclude that neither accounts for the periodic changes observed and that these must be due to a wave of assimilates passing down from the leaves in the afternoon. We may accept the distinction that xylem tensions are transmitted to the phloem less readily in *Quercus* than in *Salix*. But, though it is likely that such a wave of sugar concentration is moving through the bark, their own data suggest that not all the concentration changes can be ascribed to this cause. Fig. 8.4 contains a hint of the trouble: the sudden fall of concentration in the early morning; and in some of their other figures like that which is our Fig. 8.5, the change is quite startling in its size and suddenness. Hüber *et al.* use this figure as an example of how the concentration rise at the top of the tree may be completed while lower down the sugar is still rising, but far more striking is the abrupt fall in concentration at all heights in the early morning. Examination of their other curves, and the similar ones of Ziegler (our Fig. 3.1) show this fall to be a very general

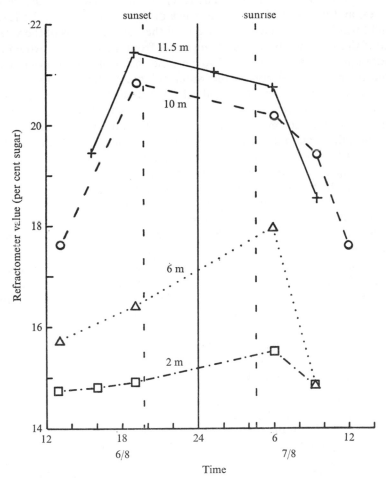

Fig. 8.5. The sudden change in concentration of exuding sap at sunrise at all heights on another tree. Re-drawn from Fig. 5 of Hüber *et al.* (1937).

phenomenon. It is suggested that there is some large change going on in the water content of the phloem which is not comprehended by the factors so far considered, and which is imposed upon the assimilate wave to the confusion of the simple interpretation; that this change is working in reverse in the evening and emphasising the rise in sugar concentration at that time.

A likely explanation of this change has been suggested to me by Dr G. C. Evans. So sudden a change cannot be due to the operation of light or temperature changes on the trunk and roots of so considerable a mass as a forty-foot oak tree. The receptors of the change must be the leaves, and the most likely consequence of the rising day that the leaves might mediate is an increase in tension of the water columns of the xylem. But if this operated by withdrawing water from the phloem it would not cause a *fall* in concentration. However, the shrinkage of the xylem mass (due to increased tension in the water columns) which is known from dendrometer measurements would release a tissue pressure on the inside of the phloem, decreasing the phloem water potential. Then the phloem cells would take up more water and the sap concentration would fall. Further, the larger fall at the top of the tree compared with the bottom, seen clearly in Fig. 8.5, would be expected since the xylem mass at the base is larger, and, containing a greater core of dead elements, will shrink less when the tension is applied. In the evening the easing of the xylem tension and the restoration of pressure to the phloem would be a slower process than in the morning and the adjustment of water equilibrium might go on for much of the night, to be suddenly upset the next morning. That the morning concentration drop is more striking in some records than others may be due to the varying weather conditions at sunrise giving different saturation deficits. It is not intended to deny that a daily wave of sugar passes into the trunk at the top; it is only suggested that the observed periodicity is more complicated than it looks.

The two other reports that this wave of sugar appears in exudates, that of Ziegler shown in Fig. 3.1, and that of Zimmermann (1958*b*), re-inforce the suspicion that distortions of the simple wave limit the use that can be made of it. Thus Zimmermann:

It is very important to choose good trees for these experiments. They should be as tall as possible with a diameter at breast height of at least 30 cm. They should have no branches on the trunk and the live crown should be symmetrical. In other words, the leaves should all be at about the same distance from the ground. This is the type of tree that grows in a forest under conditions of rigorous competition. But even 'good' trees are not ideal in this respect, and the results are therefore more or less disturbed. Further disturbances are brought about by the fact that not all the leaves get the same amount of light during the day, that the activity of the removal mechanism may be different around the trunk, and finally that an osmotic dilution of the sieve-tube exudate takes place upon tapping.

Compared with the amount of work put into the translocation rate experiments, the results were rather disappointing. An exceptionally good curve is shown ... obtained by averaging the values of two experiments, which decreases the effect of several disturbing elements.

If there is little firm ground in the analysis of exudate concentration, the kinetics of the tapped sugar can be approached in another way by following the changes of labelled sugar translocated through the piece of phloem that is yielding the sap. Then the comparison is being made independently of the water content, since radioactivity can be expressed per unit sugar mass or per unit time. When ^{14}C is incorporated by photosynthesis into a leaf on a side shoot of a rooted willow cutting it travels down the side shoot to the main stem, and spreads both up and down the bark in a narrow vertical band. The changing radioactivity of the exudate may be followed in successive samples of sap from an aphid stylet in this band

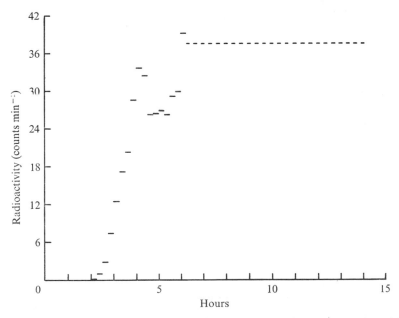

Fig. 8.6. The changing pattern of radioactivity (counts per minute per 15 minutes) in samples of sap from an exuding aphid stylet on willow bark, as labelled sugar, synthesised into a leaf 8.4 cm away at zero time, is translocated past the stylet. The dotted line represents a large volume of sap collected over seven hours. Re-drawn from Canny (1961) Fig. 1.

of bark (Canny, 1961, and Fig. 8.6). The precise form of this curve does not matter for the moment, only that it rises to a peak and falls to a steady level, and that it is mirrored by a profile of the same shape in the band of bark where the distribution of radioactivity with distance is as shown in Fig. 8.7. The changing pattern at the stylet is obviously produced by the passage of the distance profile past the observation site. Further, if the position of the distance profile is investigated just when the radioactivity of the time profile has reached its peak, the peak of the distance profile is found to lie in the segment containing the stylet. This is an important result, implying that the radioactivity measured in the stylet exudate is a good sample of the broader band of radioactivity advancing in the bark: that equilibration is rapid between the sampled sap and the transport channel wherever this may be; that there is little or no damage

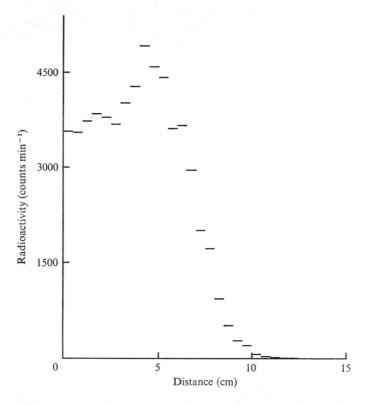

Fig. 8.7. Distribution of labelled photosynthate in the bark of a willow cutting downwards from the application shoot (inserted at zero distance) five hours after a 4-min presentation of $^{14}CO_2$ to a leaf of the side shoot. Re-drawn from Canny (1961) Fig. 3.

to the transport system by the aphid's injecting saliva, or by the act of puncturing; and that the label arriving at the punctured sieve tube is coming neither faster nor slower than the label travelling in the rest of the bark. In the paper quoted this was taken to mean that the *sugar* sample and flow from the stylet were also normal – an unwarranted deduction from the evidence. Peel and Weatherley (1962) have taken issue with this deduction, rightly: 'He (Canny) believes that the stylet gives a good sample of the sieve-tube sap in the locality of the puncture. Our view is that the stylet creates a sink in an individual sieve element which demonstrably induces movement from a distance, principally in that sieve tube alone.' The apparent paradox, normal labelled sugar flow and abnormal sugar flow from the stylet, is readily resolved by recalling the argument in Chapter 5 that labelled molecules behave in a diffusion-type system like foreign molecules. If we trust the diffusion analogy, the rate at which sugar flows from the aphid stylet will be irrelevant to the rate at which labelled sugar flows. The mass transfer of sugar down the bark may be enormous or zero; the tracer sugar will advance at its own ordered pace. The existence of the paradox is in fact strong evidence in favour of the diffusion-type system, and against a flow-type, and is a fine example of the confusion that has arisen by our trusting that the behaviour of labelled sugar in the translocation system was evidence of what the unlabelled sugar was doing.

Turning from the front of radioactivity to the plateau behind the peak, this is found to be impossible to eliminate.* We cannot produce a pulse of radioactivity in the exudate with little or no labelled sugar behind it. However short the presentation of the leaf to $^{14}CO_2$, or if the source leaf is detached after a period, or if a ring is cut in the bark between the leaf and the stylet, the specific activity of the exudate does not drop to a low value. In the latter case it drops a little, indicating, as Peel and Weatherley show, a switch of sources of sugar, but there is still substantial activity in the exudate. This must mean that once labelled sugar has passed through the transport channel it occupies a substantial space out of which it exchanges over a period as unlabelled sugar follows through. This space, necessarily rich in sucrose, could be inside or outside the channel, but the lack of any considerable pool of sucrose outside the conducting cells, and radiographs showing the label confined in sieve elements incline me to the belief that it is within these cells. The relative rapidity of the filling of this space as the profile of Fig. 8.7 advances is consistent with the slow emptying as it proceeds, because in the former case we study the filling

---

* Except in the sugar-beet leaf, see pulse chase experiments of Geiger and Swanson (1965*a*).

of the space at a point, while in the latter, label must be swept from the whole length of the channel between the source of label and the stylet.

The concept of different sources of sugar capable of supplying the sink of the exuding stylet is a valuable development of these studies. Added to the switch indicated by the girdle between label-source and aphid, is the earlier idea of a 'contributory length' of sieve tube (Weatherley, Peel and Hill, 1959), based on the observation that stylet flow was not affected by an incision in the bark further away than 8 cm on either side along the sieve-tube line. Apparently this length of stem contained enough sugar source to make up the full supply to the aphid sink when the leaf and other stem sources along the line were disconnected. The exuding stylet is seen as an efficient sink towards which sugar is transported in the tapped sieve tube from both directions, drawing on whatever sources, leaves or storage parenchyma, are situated on the sieve-tube line. Water probably comes radially from the xylem throughout this length to make up what is lost in the exudate flow. The rate of sugar arrival in the tapped sieve-tube element from one direction, is half the rate of mass flow of sugar out of the stylet. Using the figures of Weatherley *et al.* (1959) already quoted, and assuming a 10 per cent sucrose solution, this represents half of 0.1 mg sucrose per hour per 414 μm^2 or 12 g hr^{-1} cm^{-2} sieve tube, or (taking the usual figure of one fifth for the fraction of phloem occupied by sieve tubes) 2.5 g hr^{-1} cm^{-2} phloem. On the basis used for calculation of Table 3.1, the steepened gradient required to bring about such transfer would be 1×10^{-2} g cm^{-4} = 2.9 M sucrose per metre. We have an approximate check on the likelihood of this gradient from the 'contributory length' concept. The concentration ought to fall from 10 per cent to 0 per cent sugar over this 8 cm, giving 1.25 per cent sugar per cm = 3.6 M sucrose per m, the right order of magnitude. Such a picture of the functioning of the sieve tube should be contrasted with the pressure-flow–velocity calculation made above from the same data in accordance with equation 1.1. The issue is clear, but the experiments to decide between the two views have yet to be done. On the pressure-flow view measurements of the mass transfer of sugar should tie up with measurements of the area of the channel, the velocity of flow and the concentration; on the diffusion view, the mass transfer should be the product of the translocation coefficient and the sugar gradient. The difficulties hindering an experimental decision centre around the concept and measurement of the speed of flow which will be treated in a later chapter.

It is now convenient to return to some of the evidence concerning bi-directional movement which relies on evidence from aphid exudates. In this willow system just discussed the severed stylet is implanted in a single sieve element and the solute coming from it must be coming from that element. Since it did not make any difference whether label was fed into the system above or below the exuding stylet, but it always appeared in the exudate from the one sieve tube, the occurrence of movement in two directions in the sieve tube seemed indicated. Eschrich (1967) has developed this system to show that such double movement does occur. He used whole aphids feeding on *Vicia* and collected the honeydew excreted on a turntable of filter paper attached to a clock. The successive drops of honeydew falling around the circumference of the paper gave separate samples timed by their position, in what he called a 'honeydew chronogram'. ^{14}C was assimilated into the leaf of the node above the aphids and simultaneously fluorescein was fed into a leaf at the second node below (on the same vertical line as the leaf above). The drops of honeydew contained one or other or both tracers. The exudate was double-labelled in 42 per cent of the collections. Eschrich goes to elaborate lengths to explain how this might happen by other means than bi-directional movement in the one sieve tube. It seems simpler to take the evidence at its face value as supporting the other indications that the phenomenon is real.

In summary, the exudate coming from a cut aphid stylet is probably the contents of the sieve-element lumen. The long continuance of such exudation suggests that the transport system has not been inactivated by the aphid's inserting the stylet, nor, in favourable cases, by the pressure release on cutting the stylet, but that sugar continues to arrive at the tapped sieve element from sources in both directions by the normal transport mechanism. This sugar flow is abnormal in that a new sink has been established where none was before, and abnormally fast down a steepened gradient, but the advance of tracer past the new sink (revealing the speed component) is unaffected by the tapping. That the transport system should continue to function is of the utmost advantage to the aphid. Exudates obtained from incisions in the bark are much more likely to be artifacts. That solution is almost certainly more dilute than the sieve-tube sap; the drops may be sieve-tube contents forced out under pressure release on cutting; the transport system is probably damaged, since sugar does not continue to arrive at the cut, except in a few trees with very wide sieve pores, and there the flow is still relatively short. The solutes present in the exudates are likely to be food materials in transit in the translocation system, but the water content of the exudate is subject to a number of influences and concentration comparisons must be made with caution. The fragility of the translocation system – the destruction by cuts – is probably a principal reason for the limited use man has made of translocate as food; and the evolution by aphids of a tapping method that did not destroy the system has allowed them to exploit the source to the full. Further consideration of this fragility will be given in the next chapter.

9. The turgor of sieve elements and their fragility

... of the uneasy effect of the tucker of *Titian's Mistress,* bursting with the full treasure it contains.

Hazlitt

If the sieve tubes contain (as we have inferred from what they yield when pierced) a sucrose solution of up to 30 per cent (*c.* 1 M), their osmotic qualities and behaviour may be expected to require special attention. It is not as though they were the familiar balloon-type of parenchyma cell inflated by turgor against the elasticity of the walls. The sieve tubes are long, narrow segmented tubes specialised in some way for facilitated conduction lengthwise, and the question at once arises whether each segment or sieve element develops its own turgor, contained within its own semipermeable boundary, or whether a file of sieve elements shares a common turgor, enveloped by a single semipermeable boundary:

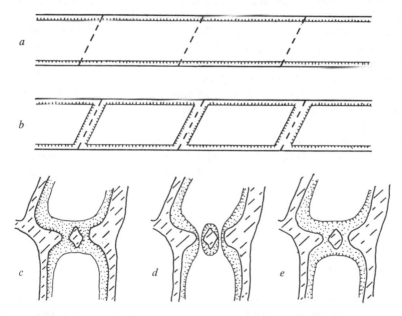

Fig. 9.1. Diagrams of possible positions of the semi-permeable membranes of sieve tubes confining the osmotic contents. *a* and *b*: longitudinal views of sieve tubes; *c–e*: details of sieve-plate pores. Protoplasm is shown dotted, cell walls hatched, and semi-permeable membranes, ⊔⊔⊔⊔⊔⊔.

whether, in fact, the semipermeable boundary forms, at the sieve plate, an osmotic separation reinforcing the visible spatial separation of sieve elements. The possibilities are represented in Fig. 9.1.

The probabilities seem nicely balanced: the facility with which solutes move along the tube suggests the minimum resistance at the sieve plates and the absence of a semipermeable barrier as in Fig. 9.1a; but the presence of a wall structure between consecutive elements suggests that the barrier should be there, lining the wall as in other cells. There is no question of the barrier passing literally outside the sieve plates for these are plainly part of the wall. The barrier in Fig. 9.1a would have to pass through and line the pores of the sieve plate, in the manner of c or d. The type-b structure requires that the barrier in some way seals off the connecting strands of protoplasm so that they fall outside the two turgor systems of the two sieve elements – a state of affairs difficult to visualise unless the semipermeable layer is within the protoplast at the tonoplast boundary (9.1c). The nature, form and arrangement of sieve-element contents are matters of much dispute as we shall see in Chapter 10, but there is wide agreement that, as the sieve element develops from a cambial initial to its final elongated form and large size, the contents change from the vacuolated protoplasmic sac, towards some diffuse gelatinous matrix by the vanishing of the tonoplast. It is clear that early in this sequence there will be at least one semipermeable layer at the tonoplast, but later, if the sieve element is still semipermeable, this property must reside in its outer membrane. While the tonoplast is still there, if the pores of the sieve plate are open, either the vacuoles of consecutive elements must be continuous through the pores (9.1d) or must round off over the surface of the plate (e). It might be thought that a few simple plasmolysis experiments would settle the question, for on the one view it ought to be possible to plasmolyse sieve elements independently in hypertonic solutions; on the other, consecutive elements ought all to plasmolyse simultaneously. But this, like other questions of sieve-tube physiology, has become involved in the polemic surrounding the question of mechanism, for the mass-flow hypothesis requires a turgor system of type a, the protoplasmic circulation theory, one of type b. Nor have technical problems been wanting.

Plasmolysis of petiole cells in an attempt to stop translocation out of the leaf was introduced by Czapek (1897). It had no effect, due, says Deleano (1911), to the fact that the 5 per cent potassium nitrate used as plasmolyticum did not plasmolyse the sieve tubes. The development of the point of view that sieve tubes *could* not be plasmolysed is interesting and instructive to trace. It seems to have grown up in Crafts's mind in the 1930s as a necessary adjunct to his firm belief in flow of solution through the sieve tubes. If the sieve elements carried a stream of flowing solution, what structure they contained must be permeable to the solute as well as the solvent, therefore they should be impossible to plasmolyse. The question of just how a flowing solution could be contained at all in cells that were permeable to it was never properly faced. There was no wall cement like the lignin of xylem elements which confined the water of the transpiration stream, but in some way the living cells bordering on the phloem might prevent lateral leakage. Certainly the sieve elements were difficult to plasmolyse. There was not much evidence on the positive side at this time, and what there was, Crafts was able to brush aside.

Phillis and Mason (1936, Further studies, VI) object that the highly permeable condition of the mature sieve tube may be an artifact resulting from sectioning. More recently, Mason and Phillis (1937) state: 'In order to insure plasmolysis of the sieve tube it is necessary to carry out the operation on the bark *before it is cut* and removed from the plant.' They do not explain how they arrived at this conclusion; and, since the necessary microscopic observation would require some dissection, one cannot easily understand how the work could be done critically. To be certain of his observations one should carry out the plasmolysis under the microscope, where the condition of the sieve tube before and after treatment could be observed. Only in this way can one determine the ontogenetic state of the tube, its freedom from injury, and its reaction to the treatment. (Crafts, 1938.)

Young sieve tubes, he admitted, could be plasmolysed, but as they developed towards 'maturity', they lost this susceptibility, along with their nuclei and much visible internal organisation:

Young sieve-tube elements are spheroid or cylindrical and well rounded. Having nuclei, slime bodies, and cytoplasm in a high state of activity, they react to stains and plasmolyzing solutions as normal vacuolated living cells. Attending the nuclear disintegration, loss of slime bodies, and alteration of permeability that occurs with maturity, are morphological changes of profound significance. At maturity the side walls of sieve tubes are slowly pressed in by surrounding cells so that the elements become funnel-shaped at each end, the transverse area in the center becoming progressively smaller. Finally the walls are pressed together, and the lumen disappears completely; the only structures left at obliteration are the sieve plates, often found lodged between the phloem parenchyma cells. (*Ibid.*)

This cloudy use of the word 'maturity' obscures our ignorance of the stage in this cycle of development and decay during which a sieve tube may translocate. The visible changes were clear, the physiological changes of function unknown; and Crafts named those sieve tubes 'mature' which were on the verge of decay, i.e. which he could not plasmolyse. He implied that they were at this stage also functionally mature, for which he had no evidence. Any sieve tubes which some other worker could plasmolyse were, in Crafts's view, immature, i.e. also non-functioning.

Schumacher (1939) tried to resolve this maturity debate by demonstrating plasmolysis in sieve tubes of leaf veins and petioles where, he argued, since there was no secondary differentiation of phloem from a cambium, the sieve tubes must be mature in both the senses of structurally developed and physiologically functioning. There was no doubt that leaf-vein sieve tubes must be able to translocate carbohydrates at least away from the leaf, and were those in which Schumacher had followed fluorescein movement. Traditional plasmolysis forms were recorded with 1.5 to 2 M glucose as plasmolyticum, and it seemed that the association of exceptional permeability to solutes with functional maturity had been broken. Rouschal, Hüber and Curtis were all involved in this controversy whose history is carefully traced by Esau (1950).

Currier, Esau and Cheadle (1955) finally settled the question in favour of plasmolysis by extensive and careful observations on a wide range of stem phloem. Their new and important contribution was the careful handling and preparation necessary for phloem tissue. Instead of the old rough hand sectioning in tap water, every effort was made to protect the phloem cells from mechanical and osmotic shock. Stem pieces were cut much longer than needed from the plant and immediately immersed in buffered sucrose solutions where they remained for less than two hours. The ½-inch segments for sectioning were cut from the centre of the large lengths by a jeweller's saw and trimmed for the microtome, being kept wet all the while with the buffer. Tangential sections about 50 μm thick were cut on a sliding microtome, with the stem, knife and sections all flooded with buffer, and the sections were mounted in buffers of different osmotic values for observation. With these precautions* plasmolysis of sieve elements was easily observed in all except one of 23 species tested. Most could be plasmolysed in solutions of 0.5 to 1 M, but sometimes strengths of as much as 2 M were needed for *Vitis*. True plasmolysis, dependent on differential permeability, was distinguished from an injury contraction by deplasmolysing again in hypotonic solution. Cycles of plasmolysis and deplasmolysis were repeated up to seven times in *Vitis*. The state of the sieve elements was precarious, and the ability to plasmolyse easily lost:

Constant comparison of the plasmolytic behavior of parenchymatous phloem cells with that of sieve elements revealed a great instability of the sieve-element

* I have called them new, but Lecomte (1889) says: 'Nous avons eu recours pour nos recherches à des matériaux frais plongés dans de l'eau sucrée à 3–5 pour 100. Les fragments à utiliser étaient détachés dans l'eau sucrée et portés dans le microtome. Le rasoir était plongé et les coupes observées dans la même solution sucrée.'

protoplast. Plasmolysis in mature sieve elements is characterised by a sensitivity to injury that requires special precautions in handling the material. A delay of one hour or more between cutting of sections and subsequent observation of them decreased the number of plasmolyzable cells. Hypotonic and hypertonic injury, from use of excessively strong sugar solutions and too dilute solutions for plasmolysis and deplasmolysis, hastened the loss of differential permeability of the protoplast. In active phloem, the surface layer of the cytoplasm which separates from the cell wall in plasmolyzed sieve elements is delicate, unrefractive, and at times almost undetectable. Companion cells invariably assumed a concave plasmolysis form, and in some species strong contraction produced a ladder-like pattern. Phloem parenchyma exhibited concave or less often convex patterns, and in some species the marked refraction of the cytoplasmic surface in these cells contrasted sharply with the delicately thin surface of the sieve-element protoplast.

And of the injured state:

Injured sieve elements were identified by failure to plasmolyze or deplasmolyze, by the formation of shrunken protoplasts with contracted lateral cytoplasm, and by the appearance of longitudinally oriented strands usually attached to sieve plates. In many tests an element would apparently plasmolyze but on the addition of hypotonic solution the protoplast would continue to shrink, thus developing a condition that we have termed *injury contraction.*... this configuration is not in any sense a normal one, but is definitely an artifact, an injury response accompanied by a loss of differential permeability. Of course, it could be that the initial separation is a true plasmolysis that becomes irreversible because the protoplast is progressively injured. Thus plasmolysis and injury contraction may have the same initial appearance and, therefore, it is not always possible to decide whether an element is plasmolyzed or not except by attempting deplasmolysis.

It was becoming clear that the sieve-tube protoplast was much more sensitive to damage than had been thought; that Crafts, who could write in 1938: 'There is at present no reason for assigning a special sensitivity to mature sieve tubes', was being much too heavy-handed; and that Mason and Phillis whom he scornfully quoted in the passage on page 117 had a truer appreciation of the delicacy of the problem. There was every reason to believe that the mature sieve element had a special sensitivity. Even when Currier *et al.* had prepared their phloem sections as carefully as they could the property of plasmolysability lasted only a little time and often did not survive at all. Several other injury reactions were easily induced and readily recognised: breakdown of plastids with release of starch grains to give progressively stronger Brownian motion; irreversible surges of cytoplasm resulting from non-uniform turgor along a sieve tube; and contracted and fibrous appearances of cytoplasm on both lateral walls and sieve plates. It was plain that nearly all previous descriptions of sieve-element structures had been of some more or less seriously damaged cells; that the recorded forms were various degrees of artifact.

No more careful study of sieve-tube plasmolysis has been published than that of Currier *et al.* and, far from solving the question of common or individual turgor, their results have served mainly to emphasise the need for still greater care. The sieve element can tolerate a certain degree of damage and still be plasmolysed, but how far this least damage may be from the state in the intact plant is not revealed by such experiments. We can, nevertheless, on the basis of other studies make another very important statement about the damage produced in excising the phloem, namely that *isolated pieces of vascular tissue are incapable of translocation.* Short pieces of stem, petiole, bark, vascular bundles, etc., on being removed from continuity with the rest of the plant are found to have lost the power to translocate sugars, dyes and so on in the high-capacity phloem transport system. They retain the power often to move other things, like auxins, at much slower rates and polarly, but the translocation system as described in Chapter I, no longer works. To support this statement with full evidence is not easy since experimenters do not hurry to report a failure. Most workers in the field must have tried to devise some simplified translocating system which could be isolated from the other complications of the plant body and studied in single-minded calm, as one makes a suspension of mitochondria. It must be judged highly significant that after all this time no such system is known. There are at least two published statements of failure. Ziegler (1958) used the vascular bundles that may be easily pulled from the parenchyma in the hollow petiole of *Heracleum mantegazzianum* to study the respiration of xylem and phloem separately, and with this material, in lengths up to 11 cm, failed to make it translocate either labelled sugar or fluorescein. The rate of travel was the same in dead and living strands. Canny and Markus (1960) in a study directed to a similar goal but using pieces of grape phloem prepared after the careful method of Currier *et al.*, and which we have seen are sufficiently undamaged to be plasmolysable, also attempted to get the pieces to move labelled sugar and failed. What appears to be an example of the isolated segment translocating, the piece of *Pelargonium* petiole with its ends in different osmotic-pressure systems, and with fluorescent dyes applied to a scraped area in the middle (Rouschal, 1941), does not, I believe, bear careful scrutiny. Any movements of dyes in the system seem explicable in terms of water flow in cell walls or xylem under the osmotic gradient. Schumacher (1950) undermines the whole work; the system has not been used again.

If the whole plant can translocate and small pieces cannot, how large a segment retains the power? This is not a question that has a simple answer. Any cut into the phloem seems to damage the translocating machinery for a distance of centimetres on both sides of the cut. Tracer sugar may be translocated towards, but stops short of the severed end of a freshly-cut shoot. The damage may be repaired over some days and a traffic of assimilates re-established to newly-forming roots on the cutting, but this is probably by the formation of new sieve tubes rather than the recovery of old ones. An instructive account of the returning power of translocation and of the bridging of a wound in *Impatiens* phloem by freshly-differentiated sieve tubes is given by Eschrich (1953). Weatherley, Peel and Hill (1959) found that exudate from an aphid stylet on a piece of willow stem was affected by cuts across the path of translocate up to 8 cm away. We may guess then that translocation will be seriously impaired in any experimental system where cuts have been made in the phloem, for a distance of 10 cm or more on either side of the cut; and may expect that the sieve tubes will have been altered from their functional state whether they appear to be damaged or not. In this light, Bauer's (1949) description of translocated fluorescein moving across the field of the microscope (Chapter 4) takes on a new significance. He was looking sideways into a transport system still attached at the ends to source and sink. From what has been learned since of the sensitivity and fragility of the sieve tubes it seems that Bauer was extremely fortunate to see what he did.

The immediate cause of the loss of function on cutting can be guessed at. The sieve-tube sap is abnormally rich in sugar, its osmotic pressure unusually high; the turgor pressure as shown from calculations on aphid stylet exudates, up to 40 atmospheres. When the tubes are cut there will be a sudden release of pressure, a surge of sieve-tube contents towards the cut, and a disruption of whatever special organisation of the contents makes it work. This pressure surge will be transmitted up the file of sieve elements through the pores of the sieve plates, dying away with the increasing resistance until a point is reached where the surge is too small to rupture the machinery. Beyond this, translocation continues to function. One may see the evidences of this pressure surge in microscopic sections where the plates bulge always one way and plugs of sieve-element contents are piled up on the concave sides of sieve plates always at one end of the element: the end towards the first cut made in preparing the tissue. Nägeli (1861) first described this appearance in *Cucurbita* sieve tubes, though he misinterpreted its meaning. Lecomte (1889) confirmed the same effect in many species and showed how it happened. The great width of the sieve pores of *Cucurbita* makes the surge particularly striking in this genus. To quote Lecomte: 'Cet effet se fait encore sentir à 10 ou

12 centimètres dans un pétiole de *Cucurbita maxima*; il est déjà très faible à 2 centimètres de la section dans un tige de *Rubus idaeus*.'

It is appropriate to introduce at this point another important property of sieve elements which appears to be related to this pressure surge, their possession of a structural carbohydrate of unusual properties, callose. This β-1-3 glucan is found in small amounts in special locations throughout the plant kingdom, for example: in some fungal mycelia, in the trumpet hyphae of the Laminariaceae; on the sieve plates of *Macrocystis*; in pollen mother cell walls, pollen grains, and pollen tube walls; in some pits; but most abundantly and constantly associated with the sieve pores of the phloem. An extensive review of its occurrence and properties is given by Eschrich (1956). Callose is readily distinguished by unique and intense staining reactions, particularly when stained with aniline blue or resorcin blue. Even more striking is its fluorescence when thus stained and viewed in UV light (see Plates 19–21). The immediate interest of callose to the question of pressure surges and sieve-tube damage lies in the fact demonstrated by Fischer (1886) and Eschrich (1956) that little is present in fresh, active sieve tubes but that it is formed very rapidly in the sieve pores when the phloem is cut. Currier (1957) found this rapid response to cutting by deposition of callose in pits of *Allium* epidermal cells could occur in a few seconds. It seems to be a plugging material which blocks the pores when flow through them becomes too rapid.

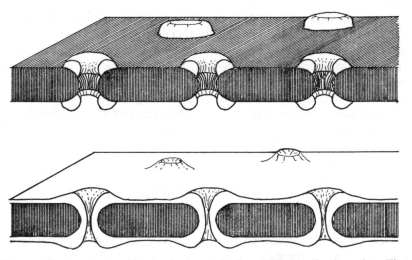

Fig. 9.2. Diagram to show the development of callose (white) on the sieve plate. First callose appears as an annulus at either end of the sieve pore. With further formation the annuli fuse within the pore to form a lining tube. The lower diagram shows the definitive callose developed over a dormant sieve plate.

Engleman (1965*a*) follows its formation with time and distance away from a wound in *Impatiens* sieve tubes. It also forms in large masses on sieve plates of dormant phloem, and disappears again when the phloem is reactivated in spring.

The plugs of callose are not simple. They are tiny annuli of new polysaccharide on each side of the sieve plate surrounding the strand that penetrates the pore, and, as they grow in size, may fuse on the surfaces of the plate and line the pores (Fig. 9.2). Even when the sieve plate is thickly coated on both sides with a callose mass, fine strands of protoplasm still penetrate the mass and connect consecutive elements through the pores (Esau, 1965). Thus though they increase resistance to flow, they may not necessarily completely stop transport if this is along the plasma connections. In some experiments callose induced locally in the phloem did not slow down the movement of ^{14}C-assimilate (Eschrich, Currier, Yamaguchi and McNairn, 1965). Indeed there is a suggestion that the translocation was faster through the callosed phloem, but this may have been caused by the boric acid used to stimulate callose formation.* However, a demonstration of the temporary blocking effect of callose is described by McNairn and Currier (1968). Maintaining a temperature of 40–45 °C over a distance of 4 cm for 15 min stopped transport and induced much callose formation. After 3–6 hours the callose went and translocation was resumed. My own attempts to repeat this experiment have been unsuccessful. Heat treatment of this intensity caused neither a blockage of assimilate flow, nor a strong build-up of callose.

Difficult though it would be to prove, the hypothesis that formation of sieve-pore callose is a leak-plugging response to wounding is an attractive one. A transport mechanism depending for its function on delicate internal organisation and containing a high pressure is vulnerable to cutting but not to bending. The plant body, being subject to bending strains at every gust of wind, could never have evolved a transport mechanism intolerant of compression. But it is cut into more rarely: by grazing animals, by storm damage or accident, and there is obvious selective advantage to the high-pressure mechanism in a response to pressure surge which constricts the channel of flow through the sieve pores. The Poiseuille law shows how efficient such a method would be: constricting the pores from 1 μm to 0.2 μm reduces the flow 600-fold. The plant does not bleed to death when the phloem is cut; transport near to the cut ceases immediately, but revives again slowly as the damage is repaired. Repair

---

* The wraith of this unhappy element, essential but of unknown function, hovers on the borders of translocation-lore. See, for example, Sisler, Dugger and Gauch (1956). The effect may operate on sugar uptake rather than translocation (Weiser and Blaney, 1964).

is probably by the differentiation of new sieve elements, and these are joined on via a special junction-cell (*Längssiebröhre* of Eschrich, 1953) to the part of the tube far enough from the cut to have resisted damage. There is a problem in establishing this junction rather like connecting up a new steam pipe without turning off the boiler, and Eschrich's description shows that the *Längssiebröhren* do not break into the high-pressure line with a new sieve plate; they are nucleated cells, connecting to the old sieve elements by narrow pits and fine but numerous plasmodesmata. Without arguing from the anatomy to the final cause we may still see that this level of anatomical organisation is entirely consistent with the view of the translocation machinery derived from physiological experiment.

We must now examine what anatomical study has revealed of the internal organisation of the sieve element.

10. The structure of sieve elements

> One also may ask why the sieve plates did not disappear during phylogeny, as in the simple perforations of vessels. This seems to indicate that sieve plates have a positive function in the transport of food material.
>
> Hüber (1957)

Because sieve elements are able to do something that no other plant cells can do, we may expect that they will have a different organisation. Because their special activity is from three to five orders of magnitude larger than in other cells, we may expect the difference to be great. Because the activity is so easily destroyed by cuts, and the properties of sieve elements so readily changed, we may expect that their special and very different organisation is going to be difficult to discover. And what we want to discover is the organisation inside the sieve elements of functioning, translocating phloem. From the final sections of the previous chapter it will be plain that this means sieve elements of intact phloem connected at one end to a source and at the other to a sink, and interfered with as little as possible from the sides. There are several different sorts of phloem tissue available for study: proto- and meta-phloem of the primary body, and secondary phloem, in stems and leaves and roots of angiosperms, of gymnosperms and of pteridophytes; but since they are all doing the same job, and, as far as can be seen, doing it about equally well, it is the similarities and common features that will be important, not the differences and singularities. There is need of a wide survey to establish the constant features, but the special ones may be left to specialists.

One common feature is that the sieve element is short-lived. It differentiates quickly from a meristematic cell (procambial strand or cambium) passing in a short time through the series of states to be described, and within a few weeks, a month or two, or (in exceptional cases in some deciduous perennials) two seasons, loses its contents, becomes squashed and obliterated.* It remains to be shown whether the sieve elements of arborescent monocotyledons remain functioning for the long life of the trunk (see Tomlinson and Zimmermann, 1967). The continued renewal of so fragile a tissue is a necessity, but makes the recognition of functioning phloem difficult (as there was occasion to note in Chapter 9) and also means that in a small region of phloem there may be sieve elements of different functional abilities. In primary phloem therefore, with non-

---

* Note in passing the high selection pressure acting to keep an expendable tissue on the outside of the expanding body.

orderly patterns of sieve elements and other cells, functional sieve tubes are more difficult to recognise, but they are easier to examine intact for mechanical and geometrical reasons. Paring away soft cortical tissue can leave a vascular bundle exposed to view and illumination while still connected at both ends, and *Heracleum* offers the experimenter such bundles already freed of surrounding tissue. Secondary phloem is produced in orderly sheets from the cambial surface, presenting a radial succession of progressively older elements from the youngest, barely recognisable as potential sieve elements, to the states of disorganisation. For this reason it has been a favourite for the study of fixed phloem. Find the cambium and you have a base line from which to measure the progress of differentiation, and distance from it gives the measure. Fresh secondary phloem can be sectioned readily on a tangent to the axis and studied microscopically, but the cuts that delimit the section have stopped it translocating, and the relevance of what it shows will always be doubtful. There are three tests available for identifying a functioning sieve element: its ability to translocate visible dye while it is being observed (applicable to any preparation); its ability to translocate radioisotope; and the presence of the stylet tip of a *feeding* aphid. The last two can be assessed only in fixed and sectioned preparations and so are useful in studying secondary phloem where the important thing to be known is the zone of functional sieve elements – the distance from the cambium.

From these last two tests it has been argued that functional sieve elements are the young ones, newly differentiated from the cambium. Kollman and Dörr (1966) selected aphids on *Juniperus* which were exuding honeydew and traced the stylets to the source of sugar. This was always the youngest recognisable sieve element in the succession away from the cambium. Kollman confirmed the finding in *Metasequoia* by the other test (1967) by showing that labelled assimilate was localised to the band of sieve elements two cells wide between the cambium and the first band of phloem fibres. On the other hand the aphid-stylet test reveals functional sieve tubes of a previous season in *Tilia* (Evert, Eschrich, Medler and Alfieri, 1968). Further, it is not certain that the aphid may select its meal on some other criterion than the presence of the translocation stream. For example, if it chooses the sap richest in nitrogen (and it excretes most of the sugar) it would prefer those sieve elements nearest the protein-synthesising centres of the cambium. It seems that to be certain of studying a functional state the youngest recognisable sieve elements should be chosen, but that those in later stages may also be functioning. An attempt is here made to set out what common, constant, peculiar features of structure can be found in sieve elements by the best available techniques of observation of both fresh and fixed tissue.

I. OBSERVATIONS OF FRESH PHLOEM

The observations of Schumacher and Bauer of movement of fluorescent dyes in functioning sieve tubes have been given at length in Chapter 4. They should be considered carefully, particularly Bauer's description, for this is the only record of what can be seen in a working sieve tube until the report of Fensom, Clattenburg, Chung, Lee and Arnold (1968). These authors also record that a dye, Janus green, moved rapidly through their preparations. Before dealing with this work it is necessary to consider the observations of fresh but detached sections, and Small's experiment. Many, from Hartig on, had examined fresh sieve elements, but with only a slowly gathering appreciation of how careful they had to be. These studies were directed to answering two questions: Could sieve elements be plasmolysed? Was there protoplasmic streaming in them? I have dealt with the first and turn now to the second.

Curtis championed the view that cyclosis streaming of sieve-element contents was the cause of the accelerated transfer even though the movement had never certainly been seen there. Streaming there was, everyone admitted, in the other cells of the phloem, especially phloem parenchyma and phloem fibres. Here cytoplasmic inclusions could be seen circulating from end to end of the cells at up to 3 cm hr^{-1}. Streaming could be clearly seen in the differentiating sieve elements of fresh sections at the stages when they were little different from other cells. But by the time they had become distinctly sieve elements, with developed sieve areas and their characteristic glassy emptiness, no streaming could be seen. Even Mason and Maskell, whose general ideas it suited best, agreed that they could not see it, and calculated (Mason, Maskell and Phillis, 1936a) that it would have to be impossibly fast to produce the rates of sugar transfer they had measured.

There was one positive statement. Small (1939) made a very careful operation on an intact *Cucurbita pepo* stem with specially sharp knives. He records seeing particles moving in both directions in sieve tubes at rates of 2.4 cm hr^{-1} or higher. One positive result cancels many failures. Because of the extreme importance of this result if it is true, it is worth quoting Small's whole account.

The explanation of the very high rate of translocation of carbohydrates is one of the outstanding problems of plant physiology.

In a successful endeavour to observe streaming of protoplasm, or at least movement of particles, in mature sieve tubes the writer has devised a delicate technique with which it may be possible to make extended observations. Meanwhile he is of the opinion that, with the publication of the technique, the field of experimenters might be considerably widened.

The preparation of the tissue for observation is essentially a delicate operation. The elimination of the stoppage of streaming by shock, noted for other materials by Ewart (1903, pp. 72 sqq.), may require treatment as careful as that given to operations on the human eye. Acting upon this idea the writer had two ophthalmic lances, 2 mm wide and 2 cm long, specially made for these experiments by Messrs. Grieshaber of Schaffhausen, and also obtained from the stock of the same makers two other ophthalmic lances, 3 mm wide and 3 cm long.

Young plants of vegetable marrow were transplanted into large flower-pots, one to each pot, and tied up to canes as they increased in length. A plant in its pot was transferred from the garden to the laboratory bench. A straight internode about 5 in. long was selected in such a position on the stem that this internode could lie across the stage of a microscope without any strain. Two transverse cuts were made about half-way through the stem with a razor blade, and a single longitudinal cut was then made with the smaller Grieshaber knife. The exposed part of the stem tissue was then placed against the lower glass of a Watson's live-box. The whole stem on either side of the raised part of the live-box was then padded to fit, and bound with two rounds of tape to the brass base of the live-box. The part of the stem being operated upon was thus firmly fixed, with tap-water and saliva keeping the cut surface moist.

On the opposite side of the stem two further transverse cuts were made with the larger Grieshaber knife, rather closer than the first pair, and with the same knife a longitudinal cut was made which left a thin slice connecting the two parts of the stem. Finally, with the smaller ophthalmic lance, a single slicing cut was made, which left parts of the connecting tissue thin enough to be viewed through the microscope (about three or four cells thick). The cut surfaces were kept wet with tap water and a brush throughout these stages. Then the thin connecting strip was flooded with tap water and a cover-slip placed on top. All the longitudinal cutting had been done with the sharpest ophthalmic instruments known in Europe, on the principle that the sharper the cutting edge the smaller the shock during the operation.

Observations were made using a $\frac{1}{6}$ in. objective. Small granules in the sieve tubes were seen to be in fairly rapid movement. The movements were mainly smooth, but sometimes a larger granule would cease movement temporarily and begin again. The smaller granules appeared to move faster than the larger granules. Naturally with this successful first experiment the writer appealed to other observers. A laboratory assistant was able to check one set of movements, as to reality and apparent direction, and went on to observe other movements also. Prof. D. C. Harrison of the Department of Biochemistry, Queen's University of Belfast, also checked the reality of the movements, so that there can be no doubt that movements have been observed in the mature sieve tubes of *Cucurbita Pepo* var. vegetable marrow.

The movements occurred in quite a number of sieve tubes, and were observed in both directions, up and down the stem. The speed of movement of some of the larger granules was timed with a micrometer scale and a stop-clock. The speed varied from 0.13 to perhaps 1.0 or more mm per minute; 0.4 mm per minute was the highest speed measured; more rapid movements were observed.

The approximate speed in mm per minute for other materials are: *Nitella* 2-3, *Vallisneria* 0.7, *Elodea* 0.96 (Ewart, 1903, p. 25). From the character of the movements it may be considered that the actual speed of the streaming may be higher than these first observations would imply, but this and other points, including the question of the observation of streaming in sieve tubes of other kinds of plants, involves careful development and extension of the technique which is here described.

Reference

Ewart, A. J. (1903). *On the Physics and Physiology of Protoplasmic Streaming in Plants.*

Though we may regret that the case is not fully documented, it is difficult to see what more Small could have done and since this statement the possibility of streaming in sieve tubes has had to be admitted.

Crafts was during this period the most determined observer of fresh phloem in his attempts to establish the 'pre-mortal', highly-permeable, non-plasmolysable condition that fitted his ideas on mass flow. He showed the sieve elements had very little in the way of visible inclusions, no nucleus, no clearly delimited vacuole (Crafts, 1939*b*):

The nucleus disintegrates and disappears... the plastids dissolve releasing their granules into the central vacuole; streaming slows and gradually ceases; the cells fail to accumulate vital stains and no longer plasmolyse... the parietal cytoplasmic layer loses its clear inner phase boundary; the cell apparently becomes readily permeable to assimilates in solution permitting free flow of sap through its lumen.

What remained in the sieve element was either a diluted cytoplasm or a gelled vacuole which had as little visible structure as a jelly, and many of the properties of one.

The question of the physiological state of this permeable cytoplasm should interest cytologists for not only does it represent a unique case of adaptation, but it is the only instance where protoplasm in such a low state of activity persists so long in a functioning condition. Hüber and Rouschal (1938) term this a 'pre-mortal' or 'pre-lethal' state, implying that the protoplasm has passed through its period of highest activity and while approaching death has not yet succumbed.

On damage, operational, osmotic or autolytic, the clear gel became fibrous, resolving itself into a number of tough elastic strings clinging to the sieve plates:

From a fragile semi-fluid layer which readily shows concave plasmolysis and tears away from the end and side walls freely the cytoplasm becomes thinner, tougher and more fibroid, and eventually becomes so firmly embedded in the sieve plates that it only comes away by violent fracture. Though the cytoplasm in this state is highly permeable and will not plasmolyse, when treated with alcohol, iodine or concentrated salt or sugar solutions that dehydrate it rapidly it shrinks, pulling away from the side walls, and forms the characteristic funnel-formed sheath at each end of the cell that has been illustrated so often in text books... Micromanipulation shows this cytoplasm to be stranded in texture, tough and elastic, and entirely changed from that of the immature elements.

There was no streaming visible.

Further, Crafts points out that if there is any streaming at the rates calculated to be necessary it is going to be impossible to see with ordinary microscopy (Crafts, 1939c): 'It would be physically impossible to observe movement at 30 cm. per hour under the microscope at magnifications of 200 × or more.'

It is not plain in the context just what Crafts means by this statement, but I believe he is here stating an important fact that many have forgotten in making their observations, namely that linear magnification in the microscope produces also a magnification of observed speed. Take for convenience a linear speed of 36 cm hr^{-1}. This is $36/3600$ cm sec^{-1} or 100 μm sec^{-1}. A particle travelling at this speed will cross the field of the microscope under oil immersion in about a second. The acuity of human vision will probably not distinguish it, especially if it is small. Only by flash photography or high-speed cine techniques is it likely to be revealed. The shimmering, imprecise blur that Bauer records takes on a new significance.

Clearly the precise conditions under which Bauer made his observation are of the greatest possible interest. The material was detached petioles of *Bryonia dioica* with the laminae removed, a system which, because of its isolation, might be expected not to translocate.

To meet all objections against the method or the interpretation of the stain-images seen, I have followed directly the advancing movement in the microscope field. The technical difficulties were great, but were at last overcome.

By means of sharp small knives, produced from unused razor blades, the leaf stalk was reduced at the site of observation to a thin layer in which a vascular bundle lay. For the necessary resolution the layer must be not thicker than the phloem thickness; in substantially thinner layers I have observed no movement, perhaps because of damage, so the suggested thickness may represent about the best for observation. If the dye moves in a sieve tube near the upper surface then the microscopic fluorescence image is not different in its clarity at all from the thinnest longitudinal sections of previously-stained preparations. $1\frac{1}{2}$ cm above the shaved piece of stalk the dyestuff was applied as before and underneath the dye-wound a thick ring of vaseline was applied which prevented the movement of the dye solution. The observation site must always be kept well moistened. The prepared leaf was rapidly fastened to a 9×12 cm glass plate, and the observation site covered with a thin coverglass; distilled water was fed in under the coverglass sideways, which once more was protected from mixing with the dye solution with a dam of vaseline. The observation site was exposed to inspection with the ultraviolet light only for the short time of the main observation. Many preparations were thus produced, until just five reliable assessable movement images were obtained with K-fluorescein and berberin sulphate. Since all these experiments showed the same thing, and because of the great preparative difficulties, further experiments were abandoned. The magnification was $\times 200$, the dye concentration was $1 : 1000$.

The reader should refer back to page 54 and remind himself of what Bauer saw.

About this time also, there were observations of the appearance of fresh and functional sieve tubes recorded by Hüber and Rouschal in their attempts to settle the plasmolysis question. Hüber (1932) had tried the observation of intact sieve tubes by incident light and failed to see streaming.

The observations of sieve tubes were done with the Leitz Ultropak water immersion system which excels in depth of focus and light intensity the other direct illumination systems I have previously used. After a little practice it is very useful[1]. On account of the weak refraction of their walls, the sieve tubes always remain the darkest part of the field of view and are not so easy to investigate as epidermal structures, collenchyma and xylem. This quality however helps the observation of their contents, which now as a matter of fact not so regularly contained particles that light up as I expected from the literature, and now saw at the first examination. In addition to the well-known *Cucurbita*, observations were made on the small but clear sieve-tubes in the central vascular bundle of *Pelargonium* petiole already used by Schumacher, and also on the massive sieve-tube region of the median bundle in the sunflower petiole. For the experiment the outer cell layers had to be cleared away one by one while the tissue was kept flooded, until the sieve elements appeared. Having got so far, about three layers of sieve-tubes are accessible to microscopic investigation. Although experiments were done with entire potted plants or plants cut under water, and days were chosen as warm and favourable as possible, I have never been able to observe the faintest displacement of the sieve-tube contents in the months of September and October. I put the blame meantime on the risk of the operation; in part also on the unfavourableness of the conditions, and am convinced that some one will find the movement some day and investigate it.[2] I would be pleased also if other efforts were made in this direction. How fine it would be if we could show a photographic record as an objective document of the sieve-tube stream, in which the particles (during the duration of the exposure) should have their own proper motions within the sieve-tubes revealed like planets on an astronomer's plate!

[1] The ocular magnification was UO 11 $\times$ (oil- and water-immersion) and the

proper observation objectives: UO 50×–UO 90× (water-immersion). The use of stops and colourfilters (mostly green filters) was helpful. I thank the firm of Leitz for the generous provision of all help.

[2] How strongly dependent the study of the assimilate stream is on the realization of optimal conditions, Schumacher (1931,* pp. 789–99) has previously recorded in heartfelt sentences to which I subscribe in detail.

Besides the absence of observed streaming, note should be taken of the absence of visible contents in these sieve tubes. This empty glassy appearance was verified in the later observations (Hüber and Rouschal, 1938) when an incident-light microscope was fixed to the scraped bark of trees:

Although thanks to this arrangement the study of living tree bark *in situ* was possible for the first time, the particular observational difficulties of sieve tubes were not yet fully overcome: namely, while the parenchyma stood out clearly with all its contents and organelles, the sieve tubes seem, as the senior author said in 1932, in the incident light microscope remarkably dark and, above all, empty of contents. Even the slime body (in living tissue often highly swollen) is in no way distinguishable from the rest of the contents in clear images of the sieve tubes; only the suspension threads (Aufhängefäden) are sometimes visible. The plasma was visible only in damaged sieve tubes.

Note again the optical homogeneity, the occasional appearance of thread-like structures, and the visibility of contents being associated with damage. The more care one took and the closer one came to having a preparation of sieve tubes which might be intact enough to show streaming, the less there was to see. Interest in the practical observation of streaming in sieve tubes lapsed with the war.

In 1961 the whole question was opened again and in a new and startling form. Thaine, confident that there was still much to be learned about sieve elements from the study of fresh hand sections, produced a film of motions in phloem of fresh sections of *Primula obconica*, and published a short account of what the film seemed to show with some stills from it (Thaine, 1961). The sieve elements were seen under bright-field illumination with a small aperture diaphragm to contain longitudinal striations from one sieve plate to the other and particles were seen moving, apparently along these lines, at speeds from 3 to 5 cm hr^{-1}. The moving particles appeared not to stop at the sieve plates and re-circulate, as they would in cyclosis streaming, but to pass through the plate into the next sieve element. Thaine named the lines 'trans-cellular strands' and pictured them lying stretched from a pore in one sieve plate to a pore in the next sieve plate across the sieve-element lumen. He believed the particles were confined to these strands and moving in them, and passed through the pore into the continuation of the strand in the next element. He was able

* An error for 1930.

to follow in his preparations, though not easily film, the progress of a particle through several successive sieve plates.

'The surprising fact is that transcellular particle movement has been observed through as many as ten sieve tube elements (total length 1.0 mm) in the linear file of a single sieve tube, in spite of the damage done to transcellular strands at the cut ends of the section.' He later (1962) elaborated a hypothesis, interpreting the transfer of sugar as the carriage of sugar-rich particles by the transcellular strands.

The idea and observations were unsympathetically received by Esau, who, as the acknowledged leader in phloem anatomy for a generation, carried most opinion with her. The lines, she said, were not protoplasmic structures at all, but diffraction artifacts produced from out-of-focus images of cell walls. She showed that fixed and denatured preparations revealed similar lines (Esau, Engleman and Bisalputra, 1963). Further, it was claimed, the streaming was not in the sieve elements at all, but in parenchyma cells above or below. The moving particles appeared to cross sieve plates because in the course of their cyclosis movement in the adjacent cell, their path crossed the *image* of the sieve plate in the sieve tube. The identity of the cross walls that Thaine called sieve plates is also questioned. Esau *et al.* point out that the sieve elements of *Primula obconica* are very narrow, and the cross wall in Thaine (1961) by its size is probably a parenchyma wall.

The controversy is not settled. Thaine continues to maintain the reality of what he sees (1962, 1964*a*, *b* & *c*, 1965; Thaine, Probine and Dyer, 1967) and has not been without support (Parker, 1964*a* & *b*, 1965*a* & *b*; Evert and Murmanis, 1965), but it is unfortunately very difficult to publish convincing evidence of ephemeral microscopic motions. Several more films have been made, but even these are less convincing than actual microscopic observation. The observer's mind is able to integrate visual information with the information from the focusing fingers and build up a composite three-dimensional impression which it is impossible to convey on film. In observations in my own laboratories we have never been convinced that the motions seen should certainly be interpreted in Thaine's way rather than Esau's. It should be stressed that the novelty of Thaine's observations would lie in the movements, particularly movements through the pores, rather than strand-type fibrillar appearances which his later papers have been at pains to verify. Comparison of the photographs in Thaine *et al.* (1967) (Plate 6) with those in Crafts (1939*b*) (Plate 7) or the drawings in Crafts (1932) (Plate 8*a*) or even Fischer (1886) (Plate 8*b*) shows that there is no dispute about the occurrence of strands there. These strands are apparently a symptom of some degree of damage, possibly of less damage than makes granular inclusions visible. The 'Aufhängefäden' of Hüber and Rouschal are recalled. Strand and fibrous structures are stabilised in certain fixed images as discussed below. Even should it prove that Thaine's observations are wrong, the work will be remembered as of the first importance for the stimulus it provided for further study. In attempts to prove or disprove the reality of 'transcellular strands' laboratories all over the world turned their attention to the observation of fresh phloem sections and fixed material, and to ways of improving preparations and fixation methods. The yields from this burst of activity have been substantial.

The latest and most determined of the observational programmes arising from this stimulus has been that of Fensom and his colleagues, whose results are currently being released. Recognising the need for still greater care than had been exercised before in looking at functioning phloem, Fensom returned to the *Heracleum* petiolar bundle, dissected out from the interior of the leaf stalk, but remaining joined at both ends (Plate 9). The xylem was cut away and a strip of phloem 5–8 cells thick observed with the finest optics, and no phase contrast. The thick sections and the absence of phase are deemed important. They believe that only in such thick sections is the structure of inner sieve elements preserved, and in such thick sections the phase image is hopelessly confused by stray refractions. They did however sometimes use Nomarski illumination to heighten the contrast. In these preparations Fensom can obtain sharp images of sieve elements and sieve plates in the middle of the tissue, in optical section of about 0.7 μm, the depth of focus of the objective. Looking thus, he sees a shimmering appearance (remember Bauer) too fine and rapid to be easily resolved, but which on careful study proves to show very rapid vibratory movements of apparently tethered particles. Ordinary photography shows nothing there, but flash photography reveals the particles, around 0.5–1 μm in diameter, and successive flash pictures show them in different positions. Fensom has filmed the appearances and analysed the successive frames to show that the motions appear more extensive than Brownian motion, three to eight times the speed and amplitude that would be expected of particles of the measured size. Further, the motions are intimately involved in the physiology of transport, and die away to near-Brownian rates if interfered with by cutting the channel, by chilling upstream or by darkening the leaf for a long time. They can be stimulated rapidly into activity by feeding sugar to the starved system, or illuminating the leaf. Fensom stresses that never does he see the movement of any particle through a sieve pore. The system was very easily damaged indeed, and, on damage, goes through the successive stages: (*a*) slowing to Brownian dimensions, (*b*) aggregation of the particles near a sieve plate, (*c*) collapse of the particles and whatever they are joined to on to strand and slime-plug appearances attached to the sieve plates.

The presence of visible granules in a sieve element is usually a sign of damage. Certainly one of the first reactions to damage is the disruption of the parietal plastids and the release of the contained granules which then execute motions of a violent kind, possibly Brownian. It is not yet clear whether Fensom was observing this familiar state, or was in fact making finer observations of the empty glassy state of least damage. The observations are at present too recent to have received confirmation by other workers or to be used as a basis for theory-building.

2. OBSERVATIONS OF FIXED PHLOEM

The hardy and resilient framework of plant cell walls survived almost any treatment of chemical fixation and staining that microscopists devised, and the geometry of their form, the detailed sculpturing of the pits and thickenings, the composite chemical layering of their substance, gave the investigators so much to study that they had little concern for the protoplasts that had once been inside. There was little there but vacuole anyway, and if coagulant fixatives like ethanol, formalin/acetic/alcohol or chrome/acetic left nothing but a few wisps of stainable material within the walls, it seemed a small loss. Early electron microscopy too concentrated its attention on the walls, the run of the microfibrils, the construction of the pits. Animal cytologists had always been more concerned to preserve cytoplasm since many animal tissues contained little else and their methods of fixation were different and notoriously unsuitable for plant tissue. Their electron microscopy required fixations that kept protoplasmic matrix and inclusions as little damaged as possible and provided electron contrast. One of the most important consequences of electron microscopy has been the devising of non-coagulant fixatives that do preserve protoplasm in place without shrinkage and with less damage to particles. This new attitude to fixation, the use of buffered osmium tetroxide solutions, whose electron staining properties are more striking than its ability to preserve protein structure, and later glutaraldehyde, which does begin to do the job of a fixative, cross linking proteins, stabilising and strengthening them, was taken over by plant electron microscopists. They obtained electron images that showed cells full of contents: meristematic cells with unshrunk protoplasts, the plasmalemma appressed to the wall, rounded, turgid-seeming vacuoles and inclusions. Later, even mature parenchyma cells could be obtained with stabilised parietal protoplasm, strands of protoplasm across the vacuole, and a continuous vacuolar membrane. The importance of this for light microscopy has been slow to dawn. A science, as opposed to a cookery, of fixation techniques is only now beginning, and is just yielding its first fruits in light-microscope images that are as well preserved as electron ones, with the added possibility of substance- and enzyme-histochemistry (O'Brien and McCully, 1969).

There is little to be gained in tracing the development of views that prevailed at various stages of the application of sledgehammer fixation to the delicate architecture of the sieve elements. The images have got better, more believable, less obviously shrunk or the products of pressure-surge, but we are certainly not at the end of this progress. There is much work to be done before we are likely to be able to preserve the kind of appearance that Fensom *et al.* report from functioning sieve elements. Nevertheless, microscopy of fixed tissues has revealed much about sieve-element architecture and we must survey what can be seen there with the best modern techniques, remembering always that there is probably much more to be seen.

We may write a specification for the minimum fixation requirements that may yield an image not hopelessly misleading;

(*a*) Fixation must be done on intact, functioning phloem still connecting source and sink. There must be no cuts across the line of the phloem, and none parallel to it closer than two or three cells away.

(*b*) Fixation must be as rapid as possible (milliseconds).

(*c*) Fixation must be designed to stabilise a very tenuous and hydrated protein structure in 30 per cent sucrose solution.

Sudden freezing or boiling water may well satisfy (*a*) and (*b*) sufficiently, if we can work out what to do next. The boiling water that Fischer used in 1886 can give an electron image not very different from chemical fixation of the most advanced kind (Plate 10). Some sieve-element properties and appearances will not be subject to these cautious reservations, being robust enough to survive careful modern fixation and yield reliable images. It is only when we come to consider the state of the protoplasm that we must be careful to draw no final conclusions.

The special features of sieve elements may be conveniently brought out by describing the development of a sieve-tube element in the secondary phloem of an angiosperm. The differences that have been found in other kinds of phloem and in different groups of plants are not great, and can have little to do with the basic translocation process common to all. These differences will be treated briefly later.

The cambial initial destined to produce a sieve-tube element is indistinguishable from its fellows, long, thin-walled, rich in protoplasm, nucleate, and having a number of vacuoles. Within the cytoplasm are the usual complement of plastids, endoplasmic reticulum, ribosomes, dictyosomes, mitochondria and microtubules. The initial divides at an early stage by an axial wall to form two daughter cells, one of which will be the sieve-tube element, the other a companion cell or (by one or more cross walls at right angles to the axis) a file of companion cells intimately associated with the element. Throughout the functional life of the phloem these cells and their associated parenchyma cells and fibres will have no intercellular spaces. The walls remain tightly cemented together at the angles and do not come apart along the line of the middle lamella as commonly happens in much unlignified tissue of the plant body. They are thus isolated from the gas phase of the cortex.

The wall of the differentiating sieve-tube element is often elaborated by the deposition of irregular hyaline thickening, the nacreous wall (Esau and Cheadle, 1958). The cytoplasm, as seen in the electron microscope, becomes less dense and more disorganised than that of the companion and other nearby cells, reflecting either a real degeneration, or a greater sensitivity to the preparation methods. Plastids fail to develop lamellae, but often contain starch of a special kind that stains reddish with iodine, the so-called sieve-tube starch.

The walls where the sieve plates are going to be begin to show early specialisation. (For a detailed account of the formation of the pores see Esau, Cheadle and Risley, 1962.) From the time of the first formation of the cell plate in this cross position, the endoplasmic reticulum remains associated with the wall in patches in the regions of the plasmodesmata. Cisternae of the reticulum are seen closely appressed to the wall over a small area, and here, as the wall thickens, callose is formed amid the surrounding cellulose. The ER-callose platelet around a plasmodesma in one sieve-tube element is paired with a corresponding platelet in the next consecutive element, developing round the other end of the plasmodesma. In surface view at this time the future sieve plate is seen to be covered with many small disks of callose when suitably stained (Plate 11). As the wall becomes thicker the paired callose platelets grow in size, taking up more of the surface of the sieve plate and, in section, producing a pair of apically-oriented cones (Plate 12) which meet in the middle. The pores are broken through these callose platelets, apparently very quickly, by the removal of some of the callose to form the fully-fledged sieve tube, presumably capable from this moment of translocating the organic stream. Evert, Murmanis and Sachs (1966) believe this opening begins at the middle lamella. The pore is lined, because of the way it was made, by a sheath of callose at all times. One can see a selective advantage that the wall of the sieve plate at the pores should be made of a polysaccharide whose formation and breakdown are quickly and easily controlled by the cell, both for the act of pore formation, and for the later control of pore radius and therefore movement through the pore (r^4) in the damage responses discussed in Chapter 9.

The size of the pores, their number, and the relative area of the sieve plate they take up are important to be known for any hypothesis about how material moves through the plate. The only survey of these measurements is that of Esau and Cheadle (1959) for dicotyledons. They measured the maximum diameters of sieve pores and also the means of a large number of pores in many different species. The largest they found in *Ailanthus altissima*, 14 μm; *Tetracera* sp., 13 μm; *Cucurbita* sp., 10 μm; and *Wisteria* sp., 9 μm. The pores were often lined thickly with callose, so the diameter available for translocation of materials may be better assessed as the diameter of the stainable connecting strand within the pore. For 126 species they found the results given in Table 10.1.

TABLE 10.1   *Means of maximum and minimum diameters and mean areas of connecting strands*

	Diameter μm		Area μm^2
	Max.	Min.	
	2.11	1.18	1.76

Though the pores were small their total area approached half the area of the sieve plate and sloping of the plate often increased the area of plate, and therefore of pore, relative to the cross section of the sieve-element lumen. They list measurements of the percentage of transverse cell area occupied by connecting strands (Table 10.2). They distinguish in all measurements between pore size (including the callose lining) and strand size reported here. No special precautions were taken to prevent injury callose forming, and these values may be underestimates of the space

TABLE 10.2

	Mean area per sieve plate μm^2	Per cent of transverse cell area occupied by strands
Strands in tubes with simple sieve plates	469.6	49.3

available for transport if it has been constricted by formation of callose during section preparation. The broad picture is not altered by taking the maximum diameters without the callose.

While the sieve plate is becoming thus specialised, the protoplast loses more and more of its usual form and inclusions. The nucleus becomes disorganised, the nuclear membrane breaks and disperses, releasing one or more nucleoli which may persist or may also disperse. The element is left now without its programme control for protein synthesis and must be doomed to the short lifetime we know it to have. Speculation, but no experimental evidence, casts the business of providing what control is necessary upon the nuclei of surrounding cells, especially the companion cells.

The cytoplasm becomes further dispersed and unfixable and the boundary with the vacuole, the tonoplast, either breaks away in fragments or dissolves, leaving no clear division between cytoplasm and vacuole. To speculate whether the cell lumen is now entirely filled with cytoplasm, or has become all vacuole, or is some new phase for which a special name must be found (e.g. mictoplasm, Engleman, 1965b) seems at present wasted effort. When we know what is there and what it does a name for it will be obvious. A denser layer of cytoplasm remains, with some mitochondria, plastids with their starch, and fragments of cisternae of the endoplasmic reticulum, lining the wall. Towards the centre of the lumen, current fixation reveals a network of fibrillar, filamentous, granular or tubular elements which are not readily interpretable in terms of functional contents. They could be closely similar to the functioning contents, or they could be the disrupted remains of almost any kind of metastable system. There is a conspicuous absence of membrane material such as might delimit transcellular strands, or of particles that might show up in Fensom's film (Plates 13–16). Just what the investigator sees in the sieve elements at this stage probably depends on many factors not all under his control: the state of the phloem tissue, its immediate past history, even the time of day, the extent of operative damage before fixation, the fixatives used and the mode of applying them, perhaps the impurities in the purchased materials from which the fixatives were made. In an attempt to circumvent some of these imponderables Johnson (1968) tried to prepare freeze-etched images of sieve-tube contents. He found that most preparations were severely damaged by ice crystals, but obtained promising images, one of which is reproduced as Fig. 10.12. Its similarity to the chemical-fixed contents (Plates 13–16) is clear. Glycerol injected into the cut stems reduced ice damage, but plasmolysed the tissue and created another set of artifacts.

These processes of cytoplasmic change are probably completed at about the time that the sieve plates are broken through by the new sieve pores, though the precise comparative timing of these changes is very difficult to establish in electron microscope sections, particularly as the pore formation seems to happen suddenly. Sieve-tube elements are seen either still differentiating and joined only by plasmodesmata, or differentiated and connected by pores, but one cannot tell from a section how recently the latter have been in the former stage. Fixation and section at the moment of pore perforation is a lucky accident not yet certainly reported. Much interest and controversy has centred around what is in the pores when they are opened in the fuctioning sieve tube, and a great variety of filaments, networks, strand aggregates, granules and tubules of varying densities have been reported by workers on different tissues using different fixatives and different degrees of care to avoid pressure surge. In some preparations the pores are wide open and contain no denser aggregate of electron-dense wisps than the sieve-element lumen (Shih and Currier, 1969 and Plate 13). In others dense plugs of filamentous material seem constricted in the pore and fan out in the lumen on either side (e.g. Buvat, 1963*a* and Plate 16). For a collection of references to these images and a useful nomenclature, consult Behnke and Dörr (1967). The tendency of the contents to form fibrous strands is as conspicuous in fixed images as in fresh ones. These aggregates of filaments are surely another view of the *Aufhängefäden* of Hüber and Rouschal, and the strands in Thaine's pictures (Plate 6). The sieve tube maintains these appearances for some weeks, during at least the early part of which it is able to carry out the processes the previous nine chapters have described. It may lose its functional capacity before the contents disappear and the walls are crushed by the growth of new phloem within the stem.

The companion cell, during all these changes, has maintained a more normal reaction to fixatives, and continues to produce images like any other active parenchyma cell. Its vacuolation remains slight, its complement of ribosomes high. The walls may have their surface area increased many-fold by the elaboration of protuberances like those of the transfer cells discussed in Chapter 3. Wark (1965) called these structures 'trabeculae' in companion cells of *Pisum*.

The interconnections of the companion cells with the sieve-tube element are through modified plasmodesmata of their common wall. The fact that these are sieve areas on the sieve-element side, but not on the companion-cell side, shows that the capacity to make sieve areas is a peculiarity of the sieve element, and that sieve plates are the co-operative result of two sieve elements over their common boundary, the sieve plate. On the sieve-element side the pores are wide and single; but at the middle lamella the pore branches into a number of fine plasmodesmata through the wall of the companion cell. On the companion-cell side sacs of endo-plasmic reticulum may be connected with the plasma strand (Plate 18). Plasmatic connections of sieve elements with other cells of the phloem seem to be almost exclusively *via* the companion cells. Only very rarely are plasmodesmata found between sieve elements and parenchyma cells (Shih and Currier, 1969). This lends support to a second view of com-panion cell function, that they are the entry and exit valves for loading translocated materials into and out of the conduit, and collecting from, or distributing to other surrounding cells. The appearance of the wall protuberances in Plate 2 is consistent with the idea that the companion cell is here acting as an entry port, and that its outer collecting surface is increased to speed some surface-limited arrival from the neighbouring cells. Companion cells cannot be an indispensable part of the apparatus, for they do not accompany sieve cells of gymnosperms and pteridophytes. In conifers their place may be taken by the dense vertical cells on the margins of the phloem rays, the albuminous cells (page 33).

Some appearances in fresh and fixed images are not of general oc-currence. One such is the substance known as slime. In the much-studied *Cucurbita* sieve elements, and in those of many dicotyledons, at an early stage of the loss of structure, globules appear of dense proteinaceous material which become larger, more widespread and diffuse with further differentiation until much of the lumen is filled with this amorphous mass. It is easily coagulated by fixatives, and collects after damage at the sieve plates, forming the slime plugs which are a diagnostic feature of sieve elements in sections prepared by classical methods. It is a conspicuous feature of electron images of dicotyledon phloem, and though often absent from monocotyledons and the sieve cells of gymnosperms and pteridophytes, was seen to give rise to the strand structures of the central cavity of *Smilax* by Ervin and Evert (1967), and six species of conifer (Evert and Alfieri, 1965). The slime bodies of *Cucurbita* were thought by Buvat (1963*b*) to contain RNA, but this has not proved generally verifiable. The name slime is unfortunate. A protein component of these exceptional cells is likely to prove more active and important than the description implies. A nomenclature of more precision was introduced by Cronshaw and Esau (1968*a* & *b*), who call it P-protein (P for phloem), and distinguish four types of electron image patterns of the stuff by suffixed numbers. Though the concept arises from light microscopy, and the name seems to distinguish slime from P-protein, it may well be just an abundance of the filamentous material of electron images, coagulated and visible.

The view that the disappearance of nucleus and tonoplast is a general feature of differentiated sieve elements, though widely held, has not been without its questioners. Schmidt (1917) maintained that all sieve elements did in fact contain nuclei. They did not show in every section through an element, but serial sections of the whole cell would reveal one somewhere. Evert (personal communication) finds both nuclei and tonoplasts in some fully-differentiated elements of tree phloem, and cannot correlate their survival with any seasonal or preparative factor.

The pores of the sieve areas of gymnosperms are smaller (0.05 to 0.8 μm diameter) than those of angiosperms, and grouped together in small clusters of three to five or more which may be individually lined with callose cylinders, or, by the formation and fusion of more callose on the wall surface, appear to share a common callose cylinder (Evert and Alfieri, 1965). In the centre of the perforated wall at the position of the middle lamella each group of strands connects with an enlarged, lens-shaped cavity about twice the diameter of the strands, called the median nodule, which is filled with stainable cytoplasmic material (Kollmann and Schumacher, 1963).

On the positive side there is now good general agreement about a number of constant features of the structure of sieve elements when they have reached the mature state.

(1) Mature sieve elements have no ribosomes. In contrast to the ribosome-rich companion cells, the sieve elements show no trace of ribosome-sized granules. This is consistent with the loss of nuclear control. It does not necessarily contradict the Buvat (1963b) observation of RNA, which could be present without being organised into ribosomes. That this RNA has some significance is suggested by the observation (Bieleski, 1969) that there is rapid turnover between it and translocated ^{32}P-phosphate.

(2) Mature sieve elements retain mitochondria. These remain embedded in the parietal region of the cytoplasm, and their lamellae may be less distinct than those of the companion cells, but there is no reason to believe that they do not function.

(3) Mature sieve elements have modified plastids. These, like the mitochondria, are held close to the wall. They may contain several starch grains in dicotyledons, or, in monocotyledons a crystalline array, and few or no recognisable lamellae. The starch-containing plastids are very easily broken either during fixation, or as a slight symptom of damage in fresh cells. Their presence unruptured is evidence of (fairly) good preservation in fixed or fresh images.

(4) Mature sieve elements have no dictyosomes.

(5) Mature sieve elements may retain some cisternae of smooth endoplasmic reticulum, again close to the wall, and sometimes locally stacked half a dozen tiers high.

RECONCILIATION

Neither in observations of fresh sieve elements and studies of their empty, least-damaged state, nor in the most recent fixed images with intact parietal plastids, little callose on the pores, and no sign of pressure surge in contents piled up at the sieve plate, is it certain that the images are of the functioning structure. It seems likely that the two methods at their most refined yield images of sieve elements of comparable integrity. Unfortunately for model-building, the better the preservation the less structure there is to be seen. From all these studies of structure one gathers little instruction except that there is more to be learned: that there is some structure in sieve elements very important to their functioning, very different from other cells, and very difficult to preserve. Our under-standing of this structure is at so elementary a stage that to explain trans-location by building mechanical models on its appearances will most likely lead to confusion. Indeed much of the confusion in this field has been produced by attempts to interpret the process in terms of mistaken views of the structure.

11. The proportion of sieve elements in the phloem

Early in the first chapter, in enquiring what path area to divide the mass transfers by, the question was raised about the proportion of the phloem occupied by sieve elements. For the time, the specific mass transfer was left as $g\,hr^{-1}\,cm^{-2}$ phloem. Following the evidence of Chapter 2 that the pathway of movement was very probably the sieve tubes and the growing certainty of this from all the later evidence, it becomes pertinent to take up the mass-transfer figures again and express them per area of sieve element. Particularly this is important in the quantitative analyses of kinetics, speed and energy that are to follow; for many have assumed the factor of one fifth for this proportion, and if the mass transfers through the sieve tubes are really five times those of Table 1.1 it is important to know it. After the discussion in Chapter 10 of the structure, contents and fixation images of sieve elements, it will be somewhat clearer by what marks a sieve element may be known, and with what certainty it may be distinguished from the other cells of the phloem. This, as a general problem, turns out to be no easy matter.

The popular factor of one fifth derives from two sets of convincing (but specious) measurements of Crafts (1931a, 1933). Concerned at that time to demonstrate that the cell *walls* of the phloem could be the pathway of translocation, he measured their area in fresh hand-sections and concluded they occupied *37 per cent* of the phloem in potato-tuber stalks and Cucurbit stems (italics mine). Crafts gives tables of measurements from the *Cucurbita* stem and potato stolon of the areas, in transverse sections, of internal and external phloem, and of the sieve tubes recognisable in the phloem. He finds an average of 20 per cent of the phloem area occupied by sieve tubes. He studied fresh sections mounted in water (some stained with acid violet). He does not say what characters he used to decide which cells were sieve elements.

Others, about this time were using a much higher proportion. Mason and Maskell (see page 39) having excluded ray cells, fibres and cortical cells from their measurements of area of cotton phloem, were content to call the close-knit complex of sieve elements, companion cells and parenchyma, 'sieve-tube groups' and use the area as the denominator of their measurement. Münch (1930, p. 83) is more explicit still:

In order to convert the calculated liquid movements into speeds of flow, it is necessary, as stated, to determine the proportion of sieve-element lumina in the phloem cross section. These measurements have proved however very difficult, because sieve tubes are not always easy to distinguish in cross section from other cell types, parenchyma, cambiform cells and companion cells. We have therefore initially been satisfied with a rough estimate, and taken it that the sieve tubes make up two thirds of the phloem cross section (after excluding the rays and sclerenchyma cells).

The confident tables of Crafts carried the argument over the cautious (and German) statement of Münch, and everyone used a fifth. But Munch's reservations are the prime point. It *is* exceedingly difficult to be completely sure of every cell in a section of phloem whether it should be scored as a sieve tube or not. I venture to say that with the techniques available in 1933 it was impossible. Coagulant fixatives and paraffin embedding leave nothing of recognisable protoplasm in any cells of the phloem. The two wall-features remaining for recognition, the presence of sieve plates and companion cells, are quite inconstant. In any one section there will be a number of sieve tubes that show neither a sieve plate, nor (at that level) a companion cell. There will even be small, tapering ends of sieve elements which could be classed as companion cells. It is possible that by following all the cells of the phloem through serial sections, the sieve tubes may be recognised, but Crafts does not say he used serial sections, and my own attempts have failed. Paraffin steel-knife sections are too rough and thick to retain and reveal all the sieve plates, and the recognition of sieve plates in photographs of such sections is highly uncertain. Moreover the measurement of areas on sections as thick as 10 μm is very inaccurate. The wall thickness in particular is over-estimated because walls inclined to the optical axis show diffuse images beyond the plane of focus. Boundaries between cells are wide and unclear.

Careful fixation in glutaraldehyde/acrolein, and sectioning of glycol-methacrylate embedded tissues at 1 μm thickness with a glass knife yields phloem sections that retain much of their protoplasmic material and characters. The contents of sieve elements are destroyed much more completely than those of companion and parenchyma cells, and (when stained with toluidine blue) they can sometimes be distinguished as the empty cells among the rest that retain their protoplasm. In many tissues however, this criterion does not work because the sieve elements retain slime or other coagulated contents and are still difficult to tell from paren-chyma cells. Geiger, Saunders and Cataldo (1969) attempted a sieve-element proportion measure in sugar beet petioles by a histological procedure superior to that of Crafts (glutaraldehyde, paraffin sections,

6 μm thick, stained with tannic acid/resorcinol, not serial) and reproduced a picture of one of their sections (their Fig. 9) with cells marked as sieve elements or parenchyma. The reasons for the designations are not unequivocally plain, but the proportion they record is 29 ± 3 per cent. It is possible to be much more certain of the designation of every cell in electron images, by the criteria discussed in Chapter 10. But in such pictures only a small fraction of the phloem is studied, and the labour of assessing a proportion for the sieve elements from many sections becomes as great as that of making serial sections for light microscopy.

In the course of the study on dry-weight transfer into wheat spikelets discussed in Chapter 1, Evans *et al.* (1970) estimated the proportion of the phloem of the peduncle made up by sieve-tube lumina as $\frac{1}{3}$. They used 10-μm frozen sections stained with tannic acid, iron alum, safranin and orange G, but though they examined a number of bundles, they did not follow the sieve tubes along their length with serial sections. The difficulties of identification are not resolved, and they did not mark their picture with the cells thought to be sieve elements.

The most satisfactory method of finding which cells are sieve elements I have found to be the use of a fluorescent callose stain on serial thin glycol/methacrylate sections. After staining with aniline blue, the callose of the sieve plates shines brilliantly in the UV-microscope, and the thin resin sections retain all the fragments of sieve plate. The callose fluorescence does not fade as in paraffin or fresh sections, a great experimental convenience. As a further general (but not universal) character the sieve-element lumina are often dark and empty compared with those of the other cells. Neither of these characters nor the combination of the two is sufficient to identify with certainty all the sieve elements in a single section, but by following serial sections through several millimetres most can be recognised. It is necessary to carry the series through more than the length of one sieve element, because critical sections may be lost or folded, and their sieve plates missed. To illustrate the kind of results obtainable by this method, a section is shown in Plate 19 from a series through the phloem of a cotton petiole. Cells are marked 'S' which *at some level* in the series showed by the presence of a sieve plate that they were indeed part of sieve tubes. The cells were traced through 64 sections of 2-μm thickness, after which the continuity was lost. By this time the proportion of sieve elements was 45 per cent of the phloem, but it is considered that several more cells would have proved to be sieve elements if traced further.

Another series of transverse sections of a bicollateral bundle in the petiole of *Cucurbita ficifolia* showed that 48 per cent of the external phloem and 52 per cent of the internal phloem cell area was sieve tube (120 3-μm sections). In Plate 20 the evidence is collected from the inner phloem. Section 65 has those cells marked which showed at some level evidence of being sieve tubes and the remaining pictures of the figure show the sections which provide the evidence for some of these indentifications. For the outer phloem region of the same bundle, the composite evidence is assembled in Plate 21, and selections from the series providing some of the evidence, in Plate 22.

The potato stolon used by Crafts for his estimate has small sieve tubes which do not furnish good illustrations, and so has not been used as a pictorial example here. Nevertheless, measurements were made on a series of 320 2-μm sections on the internal phloem. The proportion of sieve tubes was 52 per cent excluding the phloem rays, and 40 per cent when the ray parenchyma was included. The *Beta* petiole bundle, whose picture Geiger *et al.* publish, has promising-appearing cells in the phloem; clear, distinct and large, but the determination of which are sieve elements even with thin glycol/methacrylate sections and callose fluorescence, is not easy. The sieve plates are mostly on longitudinal walls, rather diffuse, and easily confused with pit callose. A series of 250 2-μm sections yielded a proportion of 78 per cent.

A clear example in which there can be no argument about the proportion is provided by some tropical trees in which the sieve plates are storied and occur all at one level (Lawton and Canny, 1970). There, a single sledge microtome section stained for callose, shows nearly all the sieve plates shining brightly in the UV-microscope. There, measurements showed proportions between 54 and 74 per cent.

These few investigations cannot establish that all phloem has a high proportion of sieve elements, but I hope they show that the problem exists and that a facile assumption of 20 per cent is untrustworthy. I hope I have also shown that the determination of the proportion is important for quantitative arguments and that its measurement in any particular phloem tissue is laborious and delicate. I hope this incomplete discussion will lead some anatomist to examine the matter thoroughly in a wide range of plants, and will also lead any translocation physiologist who is making measures of the specific mass transfer to trace the phloem cells through serial sections and use a realistic sieve-tube area as the denominator.

Meanwhile for the argument it will be necessary to adopt some factor, and I elect to use the two-thirds factor of Münch. The tree species with the storied plates had proportions on either side of this, and the herbaceous tissues studied were mostly showing proportions over one half. Though no doubt each of the species listed in Table 1.1 had its own appropriate factor, the mass transfers per unit phloem area are sufficiently alike to suggest that the proportions of sieve tubes are probably not very dissimilar, and if we have so far been content with a general maximum value of the specific mass transfer of 4 g hr^{-1} cm^{-2} phloem, this may now be intensified to the proper pathway and given a value of 6 g hr^{-1} cm^{-2} sieve tube. This value will be used in the rest of the argument.

12. The relation between translocation and the respiration of phloem – energy supply

... so steht doch fest, dass der Transport in jedem Falle einen speziellen Energie-aufwand erfordert ... auch die Wanderung in den Siebröhren bzw. -zellen selbst hat energiezehrende Strukturen und Vorgänge zur Voraussetzung. Dazu gehört für alle bisher in Betracht gezogenen Transport mechanismen die Aufrechter-haltung der Semipermeabilität des Siebrohrenplasmas; bei ein 'Massenströ-mung' käme dazu wahrscheinlich noch ein spezieller Aufwand für das Offen-halten der Siebporen, bei Spreitungen ein Betrag für die Bildung und Erhaltung der Grenzflächen, bei einer Verfrachtung durch Plasmaströmung der Bedarf dieses Vorganges, bei einer unabhängigen Wanderung der einzelnen Substanzen im Plasma auf eine bisher nicht geklärte Weise ... schliesslich die Energie für die Beschleunigung der wandernden Stoffe über die Diffusionsgeschwindigkeit hinaus, in manchen Fällen auch für den Transport gegen ein Konzentrations-gefälle.

Ziegler (1958)

Transfers of dry weight so much greater than diffusion as those in Table 1.1 must require that energy is expended and work done. From an early stage physiologists have tried to find out where that energy is coming from and how it is applied to moving the sugar, whether it is energy external to the plant like the heat energy that drives the transpiration water, or internal energy from the plant's own large chemical reserves; whether it is applied to the sugar molecules alone or to a solution in water, whether as mechanical force, and whether this force is generated osmotically, electrically, chemically or by some other means. Now it is not possible to calculate directly from the result of a process how much energy the process needs. The values of Tables 1.1 and 3.1 can tell us that the process cannot be spontaneous, that work must be done to effect it, but they cannot tell us how much. The quantity of energy expended in pro-ducing a particular result depends on the way in which the result is pro-duced. You can pull a flag to the top of a flag-pole on the halyard or climb up and fix it there yourself and slide down again. The result is the same but the energy expended is different. So to make any detailed state-ments about the energy needed to achieve translocation movements, detailed assumptions have to be made about the mechanical processes involved in the movement. Nor is this possible for all models. It is quite possible to think of an explicit, useful model which would produce the desired result, but to be unable to calculate its energy needs because some of the forces and interactions of the model are not known. Also it is possible that the physical assumptions used for even a very simple model (deriving from familiar simple substances and dimensions) are invalid on the size-scale and in the special surroundings of sieve elements. So the question of energy, like those of structure and pathways, comes back to the controversy over mechanism, and a calculation of energy requirement for a proposed mechanism may sometimes be useful in deciding whether the mechanism is feasible. Mason, Maskell and Phillis (1936a) make such a calculation for the explanation of their translocation coefficient of 0.07 g cm^2 sec^{-1} (Chapter 3) in terms of the Curtis protoplasmic streaming model:

In cotton the apparent diffusion constant (cf. Mason and Maskell, 1928) is 0.07 (gm cm^2 sec. for a gradient of 1 gm per c.c. per 1 cm). There are about 20 sieve-tube units per cm. so that, on the assumption that streaming causes practically instantaneous mixing within each unit, each will, under unit gradient, differ from the next in series by a concentration of 0.05 gm per c.c. If, now, mixing takes place by upward streaming from one unit to the next through half the sieve-pores and downward streaming through the other half, the net amount passing along under unit gradient will be $\frac{1}{2}R \times X \times 0.05$ gm per square centimetre per second, where R is the linear rate through the pores and X is the fraction of the sieve-plate occupied by pores. This amount must be equal to 0.07 gm per cm^2 per second, hence $R \times X = 0.07 \div (\frac{1}{2} \times 0.05)$ cm. second = 2.8 cm second. Thus taking a maximum estimate of pore area as $\frac{1}{2}$ the total cross-section we require rates of 5.6 cm second or 336 cm per minute. Streaming at this rate seems quite impossible and, moreover, would require a tremendous rate of expenditure of energy.

The energy required for this streaming through the plate may be roughly calculated as follows. With pores 1 μ diameter and plate thickness 5 μ (1/100th of the total sieve-tube length) flow of water at 336 cm per minute through the pores would require a pressure gradient of 180 atmospheres per cm. across the plate, i.e. 180 atmospheres per metre along the whole system. The expenditure per 1 cm. cube of sieve-tube track will be 1.8×10^6 ergs or 4.39×10^{-2} cals. Since 168 cm are traversed in 1 minute, the expenditure in 24 hours will be $4.39 \times 10^{-2} \times 168 \times 60 \times 24$ cals. $= 1,060$ cals. Since 1 gm sucrose yields on combustion about 4,000 cals. this involves the respiration of at least 0.25 gm of sucrose per 1 c.c. of sieve-tube sap per day, i.e. *a 25 per cent solution of sucrose in the sieve-tube sap would be completely exhausted in one day.*

I quote this calculation at length to illustrate the method and result, and also because it is the source of a series of errors in the literature which need clearing up. Palmquist (1938) believed there was an error here, and I perpetuated the mistake (Canny, 1960a). Spanner (1962) points out our error, but indicates another. Mason et al. lost a power of 10 in their multiplication and got 1060 instead of 10,600 cals per ml of sieve-tube track per 24 hours which would require the respiration of 2.5 g of sucrose per ml per day, an impossibly high rate. The conclusion may reasonably be drawn that the system cannot work like this. Either the whole model or the assumed values are wrong.

As part of the familiar wrangle about whether functioning sieve elements were alive, active and pumping the sugar, or passive, highly-permeable and 'pre-mortal', several investigations were made of the existence of a metabolic linkage with translocation. Mason and Phillis (1936) tried excluding oxygen from the phloem, and found that provided they did so over sufficient length of stem, translocation was reduced or stopped. The translocation machinery seemed fairly resistant to the exclusion of oxygen, and behaved as though it could transport oxygen for its own use from remote parts. 'The compact nature of the phloem, and the localisation of a peroxidase, leptomin, in the sieve-tube and companion cell, prompts the suggestion that a carrier of oxygen may be a factor in the oxygenation of the sieve-tube.'

Willenbrink (1957) set out to test the effects of respiratory inhibitors on the process, arguing that if sieve-element cytoplasm was actively moving the sugar, then the source of the energy was probably the respiratory metabolism. He used Schumacher's old system. Dissecting out the central bundle from long petioles of *Pelargonium* for a length of 4–6 cm, he enclosed the conducting strand in a chamber that served to hold liquid or gaseous inhibitors and protect the bundle from drying. By these means access of the inhibitor to the phloem was made as easy as was consistent with keeping it undamaged, and the significance of a negative result rendered less doubtful. Translocation was assessed by the movements of fluorescein, nitrogen compounds, phosphorus (and, in a brief test, labelled assimilates) out of the leaf. Atmospheres of nitrogen, hydrogen and carbon monoxide around the bundle had no effects on translocation. Cyanide, however stopped the movement of all the test compounds. When it was removed the movements, at least of nitrogen and fluorescein, resumed. Arsenite, azide, iodoacetate and dinitrophenol also stopped translocation, though not reversibly. At about the same time Nelson and Gorham (1957a & b) had shown that cyanide stopped movement of labelled sugars in *Soya*, and Kendall (1955) had stopped movement of phosphorus from *Phaseolus* leaves by injecting sodium fluoride or DNP into the hollow petiole.

These experiments were open to the objection that the inhibitors were not acting directly on the transport system, but were getting up the petiole to the lamina and there stopping the loading of the translocated substances into the sieve tubes. In a roundabout experiment to show that this was what was happening Harel and Reinhold (1966) applied the inhibitor (DNP) to the cut stump of a primary leaf of *Soya* hoping that from here it would have access to the phloem but not to the application site on a trifoliate leaf higher up. Also the application leaf was removed after a few minutes so that the DNP could not act on it. This treatment, far from inhibiting the movement of labelled assimilate from the trifoliate leaf, seemed to enhance it. However, in explaining this acceleration by an inhibition of rapid upward recirculation of assimilates that get down to the roots, they destroy their own arguments. ^{14}C-Sucrose applied to a cut petiole stump (following the technique of Perkins, Nelson and Gorham, 1959) gets down the xylem to the roots and in its retranslocation upwards in the phloem *was* inhibited by the DNP. Further, if sucrose applied in this way goes to the roots and re-circulates, so also probably will the DNP, and the whole point of the local stem application to the cut stump vanishes. There was no need for them to be so devious. Simultaneously Willenbrink (1966) was publishing an account of a straightforward study that plugged the gaps in the former investigation.

The same isolated bundle-system of *Pelargonium* leaf was used, but exclusively this time with labelled assimilate transported from the lamina. The chamber around the isolated bundle was filled with HCN vapour two to three hours before [14]CO_2 was fed to the leaf. After a transport period of about two hours the leaf was divided into pieces and the segments assayed separately for [14]C. In control bundles with no cyanide the ^{14}C-assimilates passed through the petiole and beyond; with cyanide in the chamber the ^{14}C content was the same as the controls up to the chamber and was very small or zero in the chamber and beyond. By probes on the living system he showed that, with the removal of the cyanide, translocation resumed in less than 90 minutes. The proportions of different substances containing ^{14}C in the translocate were unaffected by the cyanide inhibition. They failed to get such clearly localised inhibition in *Phaseolus* where the bundle cannot be exposed by operating. The inhibitory effect of cyanide, and strangely also of a low temperature, was apparent in the 2–4 cm length of petiole above the confining chamber. In this paper Willenbrink does not describe the use of DNP, but in a later review (1968) he extends the localised inhibition by cyanide to tomato petioles and stems, and records that he can get no consistent results with DNP. He, as Harel and Reinhold did, found DNP applied locally on the petiole had its main inhibitory effect on loading in the lamina. Kendall's (1955) results, which show a clear inhibition of phosphorus translocation by DNP and sodium fluoride, need further reconciling with these statements of Willenbrink and with the work of Harel and Reinhold.

The argument that these inhibitions have been not on translocation machinery in the sieve tubes, but on the permeability barrier of the sieve-tube walls, and that their effect has been to allow a leakage out of sieve tubes where assimilates are normally confined, has also been blocked by Willenbrink in the review. By autoradiography of the operated and inhibited bundles he shows that the labelled assimilate is confined to the phloem just as in the controls, that there has been no exceptional leakage or breakdown of permeability.

The reverse of inhibition, attempts to stimulate translocation by the addition of the presumed energy-mediator ATP, have not been clearly successful. It is not an easy substance to get into cells, and the difficulties of confirming that it has got into sieve elements to have a chance to act are formidable. Ullrich (1962) applied it along with fluorescein to the leaf veins of *Pelargonium* and got a clear increase of dye transport. But separate applications of the two substances at different sites on the same vein showed no acceleration of dye movement, nor could the cyanide-inhibition of fluorescein movement be relieved by ATP. Ullrich concludes that ATP may stimulate uptake into the phloem but not transport along it. At about the same time Kursanov and Brovcenko (1961, see also Kursanov, 1961) claimed a stimulation of translocation of labelled assimilate in beet leaves. They infiltrated the leaf with ATP before feeding CO_2 and did not distinguish, as Ullrich tried to do, between an effect on uptake and an effect on translocation. The effect is not strikingly large.

Much interest centred in the 1950s round the question of the rate of respiration of the phloem since there were reports that it was very high indeed. Kursanov started the excitement with a report (Kursanov and Turkina, 1952*a* & *b*) that isolated vascular bundles of *Plantago major* had an oxygen uptake of 820 μl O_2 per gram fresh weight per hour. A figure of less than a hundred of these units would be typical of most plant tissues. They calculate that the phloem fraction of the bundles has an oxygen uptake of 5000 μl g^{-1} fw hr^{-1}. This is a good spectacular quotable figure, and was much publicised in reviews but it depends on assumptions that the other tissues are respiring very little, a mistake as it will appear. Further, they found that the oxygen uptake was stimulated slightly at these high rates by the feeding of sugar, even to only one end of the bundle. This result was especially clear with pieces of inner bark of *Caragana arborescens*. In view of what has been learned since then of the failure of isolated phloem to transport sugar, one can see in this effect rather evidence of capillary spread than translocation. The depression they got by immersing the pieces totally in sugar may well be due to restricted O_2 supply, as Willenbrink points out. Willenbrink (1957), in the paper on inhibition discussed above devotes a section to further studies of phloem respiration, using, as did the Russians, Warburg manometry on dissected bundle pieces from the petioles of his familiar *Pelargonium*, the Russians' *Plantago* and bark pieces of *Caragana arborescens*, and

Omphalodes verna, Bunias orientalis, Tussilago farfara, Primula beesiana.
The new species are all chosen for the ease with which the bundles can be
stripped out of the petiole. He expresses his results in μl O_2 uptake per
100 mg fw (presumably per hour) and also on the basis per 100 μg pro-
tein-nitrogen and 10 cm-length of petiole. The expression per fresh weight
is common to everyone who has recorded such measurements. His
results are listed for comparison in Table 12.1 and agree with the Russian
work in showing that the bundle respiration is higher than that of the
cortical tissue, if they do not agree in absolute magnitude. Cutting
bundles lengthwise raised their respiration by about a quarter, while
cutting them into short lengths depressed it. He could find in a brief
series of trials no respiration change on feeding sugar to the pieces of
tissue from three species and casts doubt on the reality of Kursanov's
figures which he believes are within the range of experimental error.

Ziegler (1958) determined to resolve some of the uncertainties inherent
in the work so far, namely the relative contributions of xylem and phloem
to the bundle respiration, and the effects of using small pieces. *Heracleum*
petiole bundles in pieces 60 cm long were divided into xylem and phloem
halves, and their respiration measured separately, both CO_2 output and
O_2 uptake, and compared with pieces of parenchyma 100 μm thick. These
values are included also in Table 12.1, with the units appropriately ad-
justed. Looking further into the question of damage, he found that the
mere action of stripping out the bundle raised the respiration of the
ground parenchyma above the value for the undamaged whole organ.
The effect is ascribed to easier O_2 access. Separating xylem and phloem
lowered the respiration of both slightly. The important new fact that came
from Ziegler's measurements was that the xylem was respiring as fast as
the phloem. Others had assumed (without testing it) that it was mostly
dead and would have an insignificant contribution of oxygen uptake.
Ziegler's figures showed this was not so, that on a fresh-weight basis it had
the same high value as the phloem. This renders Kursanov's calculations
on the *Plantago* phloem meaningless. Ziegler verified Willenbrink's
demonstration that sugar feeding, symmetrical or at a bundle end, had no
stimulating effect. Testing the tissues separately for their reactions to
cyanide inhibition he found the parenchyma respiration was unaffected,
that of the phloem was strongly inhibited, and that of the xylem most of all.

The mass of figures now accumulated made it quite clear that the vascular bundles when isolated had a much higher oxygen uptake than the
parenchyma on a fresh-weight basis by a factor of about 5–10; that Kursanov's very high figure was probably illusory; that it was not necessarily
the sieve elements of the phloem which made this oxygen uptake so high.
If phloem parenchyma cells had the same sort of potentiality for respi-

ration as xylem parenchyma, then the sieve elements need be no more active than the dead xylem vessels to account for the observed rates. It was also clear that the act of taking these tissues from the organs so increased the supply of oxygen and destroyed the translocation, that the rates did not at all reflect what went on in an intact translocating organ.

Phloem by itself in considerable quantities could be got from bark of woody plants, and Canny and Markus (1960) took advantage of the careful preparation techniques used by Currier *et al.* (1955) in their plasmolysis studies to provide phloem for respiration measurements. The values obtained are low (Table 12.1) unless the phloem slices are cultured in sugar solution, when the rates are the same as Ziegler's.

Duloy and Mercer (1961) made a study like the others discussed and their values are also listed in Table 12.1. They use these values to calculate rates of phloem or sieve-tube respiration in the same way as the Russians, and again assuming that the contribution of the xylem was slight and its parenchyma no different from the ground parenchyma. With Ziegler's work already published this was hardly justified and one must treat their calculated phloem figures with the same caution as Kursanov's. They too used separated vine phloem and produced a similar result to those now accumulating in the phloem column. They showed that both bundle and parenchyma respiration were probably using orthodox tri-carboxylic acid–cytochrome oxidase pathways and, anyway for *Cucurbita*, were equally sensitive to cyanide. Sugar feeding of the tissues (symmetrically) raised the rates by up to 80 per cent.

In the second part of Table 12.1 the respiration measures are reduced where possible to the more rational unit of $\mu l\ O_2$ consumption per mg protein nitrogen per hour. The differences between respiration of vascular tissue and parenchyma are still apparent, but in most cases are much less than on the fresh-weight basis. It seems clear that the special potency of the vascular tissue is due to its high content of protoplasm and mitochondria rather than to any unique activity of these.

Another matter these investigators were feeling towards was the substrate source of the respiration, its chemical nature and histological location, and for this reason had measured the respiratory quotients. These are included where available in the third part of Table 12.1. Broadly one may accept that the RQ is close to unity as would be characteristic of a carbohydrate substrate. Also the frequent absence of much stimulation by extraneous sugar suggests that there is usually plenty of carbohydrate in the phloem to supply the respiration for many hours. The sieve tubes, as known reservoirs rich in sucrose, are first choice for its location. There was an obvious question about this source of respiratory substrate which seemed ripe for experimental attack.

TABLE 12.1. *Respiration of vascular tissues*

Author	Tissue	Temp. °C	O₂ consumption in μl g⁻¹ fw hr⁻¹ o			
			Bundle (xylem/ phloem)	Paren-chyma	Xylem	Phloem
Kursanov & Turkina (1952a)	*Plantago*	30	820	230		
	Beta	30	572	100		
Tsao & Liu (1957)	*Beta*		462	95		
Willenbrink (1957)	*Pelargonium*	20	96	44		
	Pelargonium	20	119	36		
	Primula beesiana	20	309	52		
		20	194	23		
	Bunias	20	393	64		
	Omphalodes	20	317–411	52–72		
	Tussilago	20	189–227	34–37		
	Plantago	20	452	97		
		20	500	146		
		20	263	59		
Ziegler (1958)	*Heracleum* cyanide sensit.	25	230 —	32 +++	178 +++	220 ++
Canny & Markus (1960)	*Vitis vinifera*	20				90 *230
Duloy & Mercer (1961)	*Cucurbita pepo* cyanide sensit.	25	180 +++	63 +++		
	Apium graveolens	25	450	30		
	Vitis virgatum	25	260	70		
	Vitis vinifera	25				230 *360
Canny (1960a)	*Vitis vinifera* intact plant					†220
Ullrich (1961)	*Heracleum sphondylium* in atmosphere of N₂ for up to 20 hr					†150–3

* with sucrose added to medium
† CO_2 output, not O_2 uptake, but since RQ = 1, the comparison is valid

$_2$ consumption in $\mu l\ mg^{-1}$ protein N $^{-1}$ of:				RQ of:			
Bundle	Paren-chyma	Xylem	Phloem	Bundle	Paren-chyma	Xylem	Phloem
192	85			1.45	1.36		
167	83			1.32	1.16		
86	32.5			1.08	0.96		
				1.01	1.09		
105	144			1.01	1.03		
104	54–81			1.00	1.00		
150–180	100–140			1.00	1.00		
167	103			1.04	1.03		
103±24	33±3.5	95±23	115±33		0.98±.07	0.82±0.07	0.91±0.06
168	180			1.10–1.14			
426	263						
132	140						
							0.8
							assumed 1
							0.9–1.05

All these measurements were open to the objections already outlined and it seemed to me at that time worth making an attempt to get a measure, however inadequate, of respiration of the intact system. This could be done if the substrate for respiration were the translocated sugar itself, rather than reserves already in the phloem. If the petrol lorry was running on its own load of petrol it was a fact worth knowing. There was the possibility that carbon-labelled translocate would give rise to labelled carbon dioxide. The details of this idea should be sought in Canny (1960a) but a summary mellowed by after-knowledge has a place in this chapter, and the measurement rounds off Table 12.1. $^{14}CO_2$ was assimilated to a patch of vine leaf, and a length of the petiole was enclosed in a black plastic chamber through which a slow stream of CO_2-free air was sucked. Any CO_2 produced in the chamber by the petiole tissues was trapped as barium carbonate giving a measure of respiration of all the tissues. Any $^{14}CO_2$ produced by respiration of substances translocated from the labelled path was also trapped and its amount gave a measure of the extent of this traffic. Label within the petiole was shown to be confined to certain parts of the phloem (since part only of the leaf was labelled); therefore the ^{14}C-content of the collected gas gave a measure of $^{14}CO_2$ output from the phloem. Two firm facts were apparent: (1) the $^{14}CO_2$ content of the collected gas reached a constant level after about ten hours and stayed at this level for a couple of days; (2) during this time of steady output the ratio

$$\frac{^{14}C \text{ content of gas respired}}{^{14}C \text{ content of phloem sucrose}}$$

was constant over a wide range of the two quantities. This seemed to indicate that a regular fraction of sucrose in transit was being broken down to CO_2 on its way, and offered some hope of providing a measure of the contribution of this substrate to phloem respiration. To turn this constant into a useful respiration measure a number of further measurements and assumptions had to be made (which still seem valid), leading to a probable phloem respiration in μl CO_2 loss g^{-1} fw hr^{-1} of 226 ± 35, the mean of nine estimates. The quoted error is the standard error of the mean, a purely numerical estimate of variability. The true rightness of the figure depends on the validity of the assumptions made in calculating it. These were: (a) that the sucrose in the petiole was distributed evenly in the ground parenchyma at a concentration measured there while the balance of the total sucrose content was concentrated in the phloem at a higher concentration, (b) that the density of the phloem was close to 1, (c) that the RQ was 1. It is much more certain now than it was then that the translocated ^{14}C-sucrose is confined to the sieve tubes and that the reservoir of the ^{14}C is likely to be smaller in volume and more concentrated than the general phloem volume. It also seems clear from the fine-structure studies that though sieve elements have some mitochondria and they may be active ones, yet there are not enough to account for an exceptional respiratory activity. To enquire, as once seemed important, about the enzymic competence and capacity of sieve elements, to analyse their exudates for enzymes of glycolysis and ATP, seems now unnecessary. With a clearer knowledge of the contents of their organelles we can expect them to have these enzymes and those of the mitochondrial systems in small amounts: too small to be producing all the CO_2 measured as coming from the phloem. The cells generating the $^{14}CO_2$ collected in this experiment now seem more likely to be companion cells or other cytoplasm-rich parenchyma of the phloem. It still seems sensible to regard the whole phloem as a unit for respiratory purposes and to express our value in the old units. The average from this experiment is therefore included in Table 12.1.

This same principle of comparing respired $^{14}CO_2$ with ^{14}C-content within a translocating organ has been applied in two further publications using more sophisticated equipment. The ionisation chamber will monitor the radioactivity of a gas stream flowing through it, and further values for the ratio ^{14}C-gas/^{14}C-sugar were obtained incidentally in another experiment (Canny, 1962a) for translocate moving down the young side shoots of a rooted willow cutting. This ratio was originally given the characterising letter Q, and as it has since been the object of further investigation, needs a little discussion. The original nine values of Q from the vine experiment are set out with their mean and the standard error of the mean in Table 12.2. For experimental convenience the time of sampling the gas stream in the vine experiment was 7 hours, and this time of collection came to be incorporated in the ratio, arithmetically inconvenient though it is. The ratio compares a ^{14}C output during a period (7 hours) to the instantaneous ^{14}C content of the phloem at the end of the 7 hours when the organ is cut and extracted. The content is (reasonably) assumed constant during the 7 hours which were on the day following application and while the $^{14}CO_2$ output was steady. We have therefore.

$$Q = \frac{\text{total } ^{14}C \text{ content of gas stream from organ in 7 hours}}{\text{instantaneous } ^{14}C\text{-sucrose content of organ}}$$

Subsequent determinations of the same quantity in other plants are also listed in Table 12.2. Coulson and Peel say of their determination: 'the magnitude of the Q CO_2 value is not constant, but can vary by an order of magnitude, even between the members of a pair of willow cuttings'. It is a matter of opinion what one calls constant in biological measure-

TABLE 12.2 *Values of the ratio $Q = {}^{14}C\text{-}gas/{}^{14}C\text{-}sugar$ (For definition of Q see text)*

										Mean and s.e. of mean
Canny (1960a)										
Vitis petiole	5.1,	5.0,	4.0,	5.4,	5.9,	3.0,	3.3,	10.5,	2.9	5.0 ± 0.8
Canny (1962a)										
Salix young shoot	2.9,	4.0,	3.2							3.4 ± 0.3
Coulson & Peel (1968)										
Salix young shoot	11.3,	23.8,	4.9,	5.9,	2.5,	2.0,	4.9,	3.2,	2.9, 2.4	6.4 ± 1.2
Coulson & Peel (1968)										
Salix old stem	1.0,	2.1,	0.3,	1.7,	0.7,	7.1,	8.3,	0.1,	1.8, 0.9	2.4 ± 0.9
Canny (1962a)										
Beta vulgaris petiole	8.0,	3.5,	9.9,	10.1,	6.2,	4.7,	7.2,	8.8		7.4 ± 0.9
Acer pseudoplatanus long shoot	11.9									

ment but I should be happy to accept their mean of 6.4 ± 1.2 for the young *Salix* shoots as supporting evidence for my means of 5.0 ± 0.8 and 3.4 ± 0.3 and for the proposition that Q may be a useful constant. Their results from the old stem I would agree may be too variable to be useful. Finally the table includes some values published here for the first time on some other species. This is not the stage of the argument to speculate about what relation Q may bear to translocation or whether one would expect it to change, as clearly Coulson and Peel thought it ought, in cut tissue or tissues that are translocating fast or slowly, with little sugar or much. For the present let it stand as another fact: a (fairly) constant proportional breakdown of translocated sugar, and at least one source of the CO_2 respired by phloem.

Some further insight into the biochemistry of the energy supply may be gained from an experiment by Bieleski (1969) in which ^{32}P-phosphate was translocated away from a pumpkin leaf, and the phosphate compounds in the sieve-tube exudate of the petiole were studied. Although the main transport form of phosphorus appeared to be inorganic phosphate (80 per cent of the ^{32}P was in this form), there was rapid ^{32}P labelling of the sugar phosphates, nucleoside di- and tri-phosphates, triose-phosphate and RNA, fully comparable with the rapidity of labelling of these fractions by other plant tissues. Moreover the proportion of ^{32}P in the ATP and UTP fractions was unusually high and their ratio to the diphosphates ($16/1$) was much greater than the usual level of $3/1$. Bieleski remarks: 'This combination of a normally rapid turnover of nucleoside triphosphates, coupled with an unusually high proportion and amount, must mean that the sieve tube system is supporting an unusually active metabolism.' The presence of the word 'system' stresses that this metabolism is probably a co-operative effort of the phloem cells and is not confined to the sieve tubes whence the exudate came.

In all these records of a reasonably normal respiration there is danger of losing sight of the extraordinary property that Mason and Phillis found so interesting: that translocation was unimpeded by local exclusion of oxygen, and the suggestion that it carried its own hydrogen acceptor from better-aerated regions.

Ullrich (1961) was the only one to take up this suggestive theme in more detail. He showed that fluorescein transport continued out of a *Pelargonium* leaf when the petiole and also the lamina were held in atmospheres of nitrogen. He was the first to look for an anaerobic respiration that might be providing the energy. He used lengths of phloem strands separated from the xylem, from the petioles of *Heracleum sphondylium*, which, though smaller than *H. mantegazzianum*, shares with it the easily-separable vascular strands of petiole and stem. The respiration of the phloem pieces was measured in a Warburg manometer, under atmospheres of air, as CO_2 output and O_2 consumption, and under nitrogen as CO_2 output. The value obtained in nitrogen was very little less than the equivalent O_2 uptake in air (RQ = 1), and fully comparable with other values in Table 12.1. If the same carbohydrate consumption in air is diverted in nitrogen to a simple ethanol fermentation, the CO_2 output should drop to a third. That it does not argues either an increased carbohydrate consumption (Pasteur effect) or a different pathway and a different hydrogen acceptor.

In support of this latter alternative Ullrich recalls that the sensitivity of translocation to cyanide (presumably through a heavy-metal oxidase) is not consistent with its non-oxygen-requiring respiration. He took up the question (again foreshadowed by Mason and Phillis) of an internal hydrogen acceptor, H_2O_2 or organic peroxide, and H-transfer to it via a heavy-metal-containing peroxidase that was inhibited by cyanide. Refining Raciborski's (1898a & b) histochemical techniques, he confirmed the localisation in the phloem, particularly the young uncallosed sieve tubes, of a peroxidase that was inhibited by cyanide. It would be entirely reasonable for a tissue with a large energy requirement and no intercellular spaces to develop a respiration system using a hydrogen acceptor other than gaseous oxygen, especially a transport tissue which might well carry its hydrogen acceptor along, much as oxidised haemoglobin is carried in blood. Finally Ullrich calculates the quantity of peroxide that would be required to act as H-acceptor for a CO_2 output of 100 $\mu l\ g^{-1}$ fw hr^{-1} as about 0.6 mg $H_2O_2\ g^{-1}$ fw hr^{-1} in the petiole, and more higher up the gradient towards the oxygenated regions where the H_2O_2 would be formed.

The rate at which the phloem is consuming sugar and making energy available for all its living processes may be reckoned then as around 230 $\mu l\ O_2$ consumption g^{-1} fw hr^{-1}, or in sugar-loss units, 0.31 mg hexose g^{-1} fw hr^{-1}. The complete combustion of 1 mg of hexose yields 3.75 cal so there are produced 1.16 cal in each gram fw of phloem per hour. The phloem is acting as a unit, and though not all this sugar consumption is going on in the sieve elements, they are probably the sugar reservoir, and it is possible that a high proportion of the energy produced is fed back to run the transport. The process is not likely to be as efficient as 50 per cent, so let us round off the available energy as 0.5 cal g^{-1} fw hr^{-1}. What could it do? What is it capable of in the way of making sugar move? Clearly it could not run the Mason and Phillis machinery which required 440 cal cc^{-1} sieve tube hr^{-1}. Several calculations about processes which respiration energy might run are made by Spanner (1962) and Weatherley and Johnson (1968), though they use a higher value. A similar calculation will illustrate the use of the energy figure and also explain the limits set by structure.

It is necessary, as was said at the beginning of the chapter, to assume a mechanism, and the simplest is the laminar flow of solution through the sieve elements regarded as tubes. The example quoted from Mason, Maskell and Phillis does not make this calculation explicit. For a tube 1 cm long with radius a cm, containing a fluid of viscosity η poise, flow at velocity v cm sec^{-1} requires a pressure gradient P dynes cm^{-2} where

$$P = 8v\eta l/a^2 \tag{12.1}$$

and to convert this to atmospheres it is multiplied by 10^{-6}. Take now a sieve-tube system, whose main resistance will be seen to be the pores. Let there be 20 plates per cm; let the pores occupy half the area of the tube (see Chapter 10) and be 5 μm long. If linear flow of solution occurs at 100 cm hr^{-1} (see Chapter 13) through the lumen, flow through the pores will be at twice this rate, and we have for a metre of tube

$$v = 200/3600 \text{ cm sec}^{-1}$$
$$\eta = 1.5 \times 10^{-2} \text{ poise for 10 per cent sucrose solution}$$
$$l = 20 \times 5 \times 10^{-4} \times 100 \text{ cm}$$

so the pressure is given in atm m^{-1} by

$$8 \times 200/3600 \times 1.5 \times 10^{-2} \times 20 \times 5 \times 10^{-4} \times 100 \times 10^{-6} \times 1/a^2$$
$$= 6.7 \times 10^{-9} \times 1/a^2 \text{ atm m,}^{-1}$$

where a is in cm.

The values of P for various values of a are plotted as Fig. 12.1. It is clear: (1) That for pores wider than 2 μm radius the pressure gradient is slight; (2) that the average angiosperm sieve pore (using the extensive data of Esau and Cheadle, Chapter 10) is just at the threshold of a sudden increase in the required pressure; (3) that gymnosperm sieve pores demand very large gradients indeed. Now the energy:

In a centimetre cube of sieve-tube track let the pressure drop be P' atmospheres per sieve plate and the velocity v cm hr^{-1}. The work is done mainly at the sieve plates of which there are 20 per cm. So

$$\text{work done} = 20\,P'v \times 10^6 \text{ ergs hr}^{-1}$$
$$= \frac{P'v \times 20}{4.2 \times 10} \text{ cal hr}^{-1} \tag{12.2}$$

Taking our available energy figure of 0.5 cal cc^{-1} phloem hr^{-1} (density 1), a cubic centimetre of sieve tube may have available to it the energy of 1.5 cc of phloem (taking the $\frac{2}{3}$ proportion of sieve tubes in the phloem) or

0.75 cal hr^{-1}; so

$$P' = \frac{4.2 \times 10 \times 0.75}{v \times 20}$$

and with v through the pores 200 cm hr^{-1},

$$P' = \frac{4.2 \times 10 \times 0.75}{4000} \text{ atm per sieve plate}$$

or

$$\frac{4.2 \times 10 \times 0.75}{4000} \times 20 \times 100 \text{ atm m}^{-1} = 15.7 \text{ atm m}^{-1}.$$

That is to say, the energy expended in a piece of phloem is sufficient if efficiently applied to force sucrose solution through holes the size of those in the larger gymnosperm sieve cells to give a speed through the lumen of 100 cm hr^{-1}. It had better be; the gymnosperms are there. Though their pores are so narrow, the sieve cells are often very long, so the pressure gradient is reduced. Instead of the assumed 20 cells per cm of dicotyledons, a figure of 7 might be more appropriate.

Now the proportion of sieve tubes in the phloem is seen to be important from a new angle, that of energy supply, not merely as a denominator for the measure of mass transfer. If all the cells of the phloem are producing energy to drive the solution, the greater the non-sieve-tube fraction the greater the energy supply relative to the energy need. The proportion actually taken up with sieve tubes may be the result of a nice balance between the requirement of a large enough channel and the need to supply it with power. On the argument so far developed, narrower pores will require a larger fraction of energy-supplying parenchyma. Taking again the Crafts proportion of one fifth (Chapter 11) the calculation from equation 12.2 may take account of the respiratory energy of 5 cc of phloem for each cc of sieve-tube track and

$$P' = \frac{4.2 \times 10 \times 2.5}{4000} \text{ atm per sieve plate}$$
$$= 52.5 \text{ atm m}^{-1}$$

which would force a flow through all but the smallest gymnosperm pores.

This concept must be carefully distinguished from the Münch pressure-flow idea. This says that there is an overall pressure gradient from source to sink sufficient to force solution through the sieve elements. For gymnosperms this would have to be a gradient of 10–100 atm m^{-1} and is not observed. The present statement says that if the energy of respiring

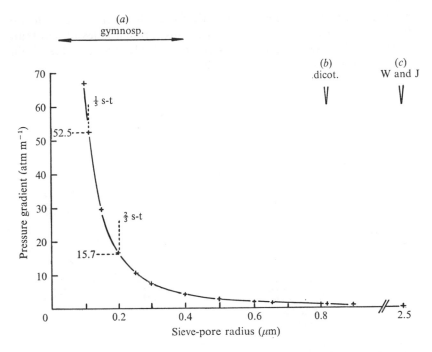

Fig. 12.1. Plot of the energy requirement (as pressure gradient in atmospheres per metre) for Poiseuille flow through sieve pores of various sizes. Calculated for a speed of 100 cm hr^{-1} through the lumen; 20 sieve plates per cm; the pores occupying half the plate. On the radius axis are marked: (a) the range of radii of gymnosperm sieve-area pores; (b) the average radius of dicotyledon sieve-plate pores found by Esau and Cheadle; and (c) the value assumed for pore radius by Weatherley and Johnson in their calculations. Also, the places are indicated at which the curve intersects the ex-perimentally-determined value of phloem respiration (0.5 cal g^{-1} fw phloem hr^{-1}) for two assumed proportions of sieve tubes, $\frac{2}{3}$ and $\frac{1}{5}$.

phloem is efficiently applied to pumping solution through sieve pores its effect can be the same as a pressure gradient of 16–50 atm m^{-1}.

However none of this decides anything about the mechanism. It mere-ly confirms that with Mason *et al.*-type assumptions the streaming in and out of pores uses more energy than is available, while a metabolically-driven mass flow through fine holes (if unobstructed) is energetically possible. Fig. 12.1 also shows that if the holes get much finer than 0.6 μm diameter by callose constriction or by being filled with the kind of cyto-plasmic filaments illustrated in Chapter 10, the energy is no longer sufficient. It should be stressed that it makes absolutely no difference how this energy is translated into pressure flow, whether by generation of osmotic, mechanical or electro-osmotic force, the energy ceases to be adequate for pushing solution by any means through holes of less than 0.4 μm diameter. Only by finding some completely different rules from the Poiseuille rules can the machinery be worked.

13. The profile of advancing tracer

When the petiole is fed to a leaf in the dark virtually no upward or downward movement of HTO occurs.

Choi and Aronoff (1966)

When you turn on the hot tap the water gets gradually, not suddenly, warmer, whereas the discontinuity at the tank between hot water inside and cold water outside was probably quite sharp. Forces in the flowing mass in the pipe have mixed some of the hot with the cold and changed a step-distribution of temperature v. distance into a smooth one. It is not at once apparent what the new distribution of temperature with distance is. Never mind for the moment. Let us call this distribution the *profile* of temperature. In a similar way when the translocation system begins transporting radioactive tracer, whatever the initial profile of tracer may be, the translocating machinery will probably change the profile progressively with distance. It may be that the way the profile changes can give us significant information about the kind of transport going on: it is certain that for experimental work we will need to know the profile and its properties, particularly before using tracers to estimate speed.

The changes in the hot water system happen even in a well-lagged pipe, with no loss of heat radially, but a possible cause for change in translocation may be sideways leakage from the channel either reversibly or irreversibly into surrounding cells. This must be considered. It will be important with 'leaky' tracers like ^{32}P-phosphate, but on experimental grounds does not seem likely to be important in the first hour of ^{14}C-assimilate advance for the following reasons: (1) Extracts of ^{14}C-assimilate in the process of translocation are almost entirely ^{14}C-sucrose, and very little ^{14}C (5 per cent) remains unextracted in the tissue. (2) The sieve tubes contain a strong solution of sucrose, and there is no nearby large, detectable pool of sucrose outside them. (3) Autoradiographs of the ^{14}C-translocate of increasingly better resolution, after ever-improving fixations, show the radioactivity more exclusively confined to the sieve tubes alone. After long times (2–3 hours), or at nodes, or in organs where there are active sinks or large pools of sucrose outside the sieve tubes, much label may be found outside the sieve tubes, but if experiments are confined to long, uniform, mature petioles or stems and to the first passage of the profile it seems probable that the profile will reflect forces operating in the pipeline, not exchanges into and out of it. Experimentally the profile will be measured either by scanning the translocating organ with an exterior probe that registers changing radioactivity with distance, or by cutting up the organ into standard lengths and assaying each. The former is non-destructive but of limited usefulness for isotopes of weak emission (^{14}C); the latter is accurate but allows a determination at only one time. Both will measure the sum of radioactivity inside and outside the channel at each distance.

I. THEORETICAL PROFILES

Horwitz (1958) considers five separate models by which the profile might develop. Briefly these are:

(1) *Flow through the pipe with irreversible loss of tracer through the walls.*
This leads to a profile in which log radioactivity (R) v. distance (x) is a straight line of characteristic slope, which at successive times moves further up the $\log R$ axis (Fig. 13.1a). Except with very rapid leakage this description does not fit the idea of the advancing profile of tracer in the sense of this chapter: the shape of the first-appearing front of the main wave of translocate.

(2) *Flow through a pipe with reversible loss through the walls.*
This leads to an equation he did not integrate. Taking the results of an analogous heat problem he draws the profiles shown in Fig. 13.2. Choi and Aronoff (1966) give a solution in terms of a series which can be summed on a computer, and provide two more curves.

(3) *Flow through a pipe with time-variable source of radioactivity.*
This is the same pipe as in (1) but the input from the source increases with time. It leads to similar lines on the semi-log plot except at short times when the profile is slightly concave downward (Fig. 13.1b).

(4) *Cyclosis-diffusion and accelerated diffusion.*
Both these lead to a profile of tracer identical with that produced by diffusion into an empty space – the non-steady-state case of Fick's law – but lead to it more quickly. The profile radioactivity v. distance is then the curve of normal error. This is important. Horwitz dismisses it curtly: 'It obviously does not conform to the known data.' It did not at that time, but it does now. Moreover, as pointed out in Chapter 5, if Mason and Maskell's data point clearly to a diffusion-analogue movement for chemical sugar, we should be prepared to accept the possibility of a diffusion-analogue behaviour for tracer sugar. This is it. This is the quantitative expression of the qualitative description of tracer movement

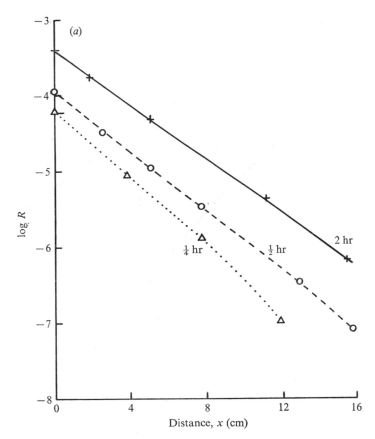

Fig. 13.1*a*. Successive profiles log *R v. x* produced by flow through a pipe with irreversible loss through the walls, at times $\frac{1}{4}$, $\frac{1}{2}$ and 2 hr of tracer movement. Speed 50 cm hr^{-1}; cross-area of pipe 0.1 mm^2; rate constant for loss 2 mm^2 hr^{-1}. Re-drawn from Horwitz (1958) Fig. 1.

in Fig. 5.2. Since it will be needed later, it may be stated now. The profile of radioactivity R with distance x away from the place of maximum radioactivity R_0 is described after time t by

$$R/R_0 = 1 - \text{erf}\left[x/2\,(Dt)^{\frac{1}{2}}\right]. \tag{13.1}$$

The 'erf' means 'the error function of'. Its values are tabulated in Appendix 2. D is the diffusion coefficient whose units are cm^2 sec^{-1}. For ordinary molecular diffusion of sugar in water its value is about 3×10^{-6} cm^2 sec.$^{-1}$. Mason and Maskell, it will be remembered, measured the coefficient for sugar in cotton translocation as 7×10^{-2} cm^2 sec^{-1} (Fig.

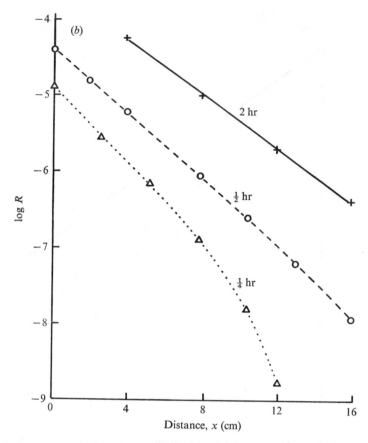

Fig. 13.1*b*. Profiles as in Fig. 1*b* but with a time-variable source of radioactivity. The shape of the lines is dominated by the leakage. Re-drawn from Horwitz (1958) Fig. 1.

3.2), and it was called in Chapter 3 the translocation coefficient, symbol K. The profiles of R v. x are shown on linear and semilogarithmic plots in Figs. 13.3 and 13.4, with $K = 7 \times 10^{-2}$, for various times. At very long times the tracer assumes the profile of the chemical gradient. Horwitz shows that diffusive spread with irreversible loss leads again to a straight line on the semi-log plot.

(5) Surface flow.

Spreading of substance at an interface between two phases, under the gradient of surface pressure, was proposed as a possible means of translocation by Van den Honert. It has found few adherents in theory or

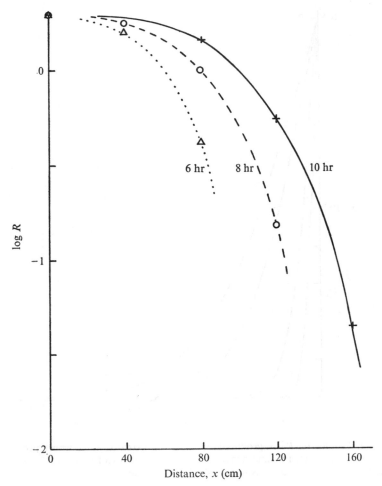

Fig. 13.2. Successive profiles of $\log R$ v. x produced by flow through a pipe with reversible loss from the walls at times 6, 8 and 10 hr of tracer movement. Speed 20 cm hr^{-1}; rate constants for both outward and inward passage 1 hr^{-1}. Re-drawn from Horwitz (1958) Fig. 2.

experiment. Horwitz shows it leads to a profile with an elaborate equation, and he does not plot it.

Now among these models Horwitz has not considered the hot-water tap. Indeed he specifically excludes any attenuation of tracer concentration with distance produced by flow alone, but it happens, and it is important. Flowing fluid behaves quite differently at low speeds when the velocity distribution is smooth and continuous (laminar flow) and at

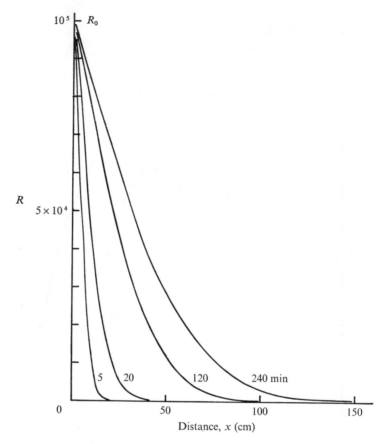

Fig. 13.3. Profiles of $R\,v.x$ produced by a diffusion-analogue transport process at successive times 5, 20, 120, and 240 min, with a coefficient (K) of 7×10^{-2} cm² sec⁻¹, equal to that measured by Mason and Maskell in cotton.

high speeds (turbulent flow). The transition between the two states is usually sudden, and is described by the size of a quantity, the Reynolds number. In a pipe, if the Reynolds number is less than about $2-4 \times 10^3$ the flow is laminar, if more it readily becomes turbulent. For a smooth pipe of radius a cm and fluid of dynamic viscosity v (0.011 for water) flowing at mean velocity U cm sec⁻¹, the Reynolds number is

$$N_R = 2aU/v.\qquad(13.2)$$

Model 6

The behaviour of soluble material dispersed in fluid flowing in pipes has

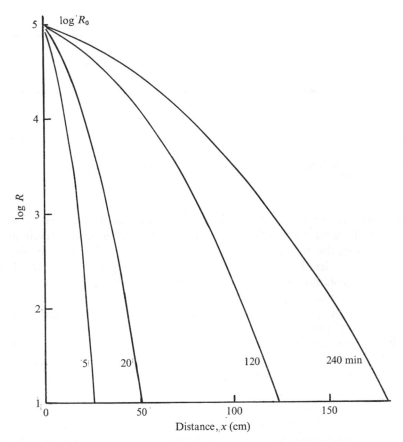

Fig. 13.4. The data of Fig. 13.3 on semilogarithmic axes, $\log R\ v.x.$

been investigated in two delightful papers by Taylor (1953, 1954). The first deals with laminar, the second with turbulent flow. The papers are so little known to biologists and the results so unexpected that a summary will be given here, but the reader is urged to consult the originals for his entertainment and instruction.

In laminar flow of liquid in a circular pipe the liquid on the walls of the pipe is stationary while in the centre of the pipe the liquid moves at maximum velocity u_0 equal to twice the mean velocity U. The distribution of velocities across a diameter is parabolic and the mean velocity U is the velocity of a meniscus where the fluid ends (Fig. 13.5).

Taylor's analysis showed that a coloured solute introduced into flowing water as a concentrated zone should spread out into a symmetrical column

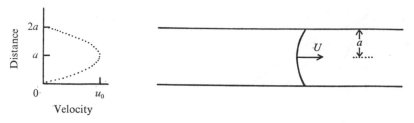

Fig. 13.5. Laminar flow in a pipe. The meniscus travels with speed U, the mean speed of the different layers at different positions in the pipe. On the left is shown the distribution of velocities across the pipe: zero at the walls and a maximum (u_0) in the centre. $u_0 = 2U$.

of colour of slowly increasing length as it moves. The centre of this column of colour moves with the mean speed U. Two problems fascinated Taylor about this result: (*a*) that the very asymmetrical forces of the parabola of velocities should produce a symmetrical dispersion, and (*b*) that though the column of colour travels at the mean speed U, the water at the centre of the tube has a velocity $2U$. Thus clear water must be catching up with the colour, passing through it and coming out clear again on the other side. He shows by experiment that these things are so.

If the time of flow is short compared with $a^2/(3.8)^2D$ (where D is the coefficient of molecular diffusion) then the profile of colour with distance is a straight line whose slope decreases as the flow proceeds. In Taylor's experiments this condition was achieved in a tube of 1 mm bore with flows of about 100 cm in $1\frac{1}{2}$ sec. If, however, the flow is long compared with the above fraction ($a = 0.025$ cm and times of up to 11,000 sec for 60 cm) molecular diffusion now contributes to the profile and the curve of colour *v.* distance is the error curve of equation 13.1. The colour is 'dispersed *relative to a plane which moves with velocity* $\frac{1}{2}u_0$ exactly as though it were being diffused by a process which obeys the same law as a molecular diffusion but with a diffusion coefficient k, where $k = a^2u_0^2/192D$'.

The mean speed is more use than the maximum so we can simplify this virtual coefficient of diffusion to

$$k = a^2U^2/48D. \tag{13.3}$$

So here we have the error curve turning up again, generated this time by slow laminar flow in a tube, and with an enhanced value of the coefficient just as Mason and Maskell found an enhanced value for the diffusion coefficient. It will not be surprising if between this effect and the accelerated diffusion analogy error curves are found in tracer profiles. The results of one of Taylor's experiments with his fitted error curves are reproduced as Fig. 13.6. The patch of colour is seen at three stages of its slow movement down the tube. The mean speed U of the advance is that of the plane of symmetry, not the advancing front of detectable colour.

Consideration of equation 13.2 for tubes the size of sieve tubes and sieve pores will show that the flow is not likely to be turbulent for any conceivable speeds of movement of sieve-tube contents. For turbulent flow, again an error curve is generated, but with a different value of the virtual diffusion coefficient. The effect on the dispersion of interrupting the flow with sieve plates can be considered qualitatively by noticing that in equation 13.3 the molecular diffusion coefficient occurs in the denominator. This is the paradox that Taylor emphasises, that diffusion tends to keep the solute together, not to disperse it. Any process that increases diffusive mixing, such as interruption with perforated septa, should further reduce the dispersion of the solute, measured by k.

Model 7. Dispersion of a profile by multiple pipes of different diameters. All the models so far have dealt with pipes of a single radius, but the anatomical reality is of tubes with different radii, and clearly the narrower

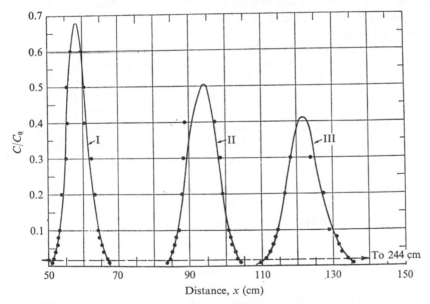

Fig. 13.6. Profiles of colour dispersed by slow flow through a tube, I after 29 minutes, II after further 36 minutes, III after further 5 minutes. Flow in a tube of radius 0.025 cm at 'a small fraction of a centimetre per second'. Reproduced by permission from Taylor (1953) Fig. 8.

ones are going to impose slower flow. Thus simple laminar flow of a sharp front of dye solution into a bundle of pipes of different radius is going to produce a dispersion of the sharp front. The form of the profile produced is most easily aproached graphically. Assume the radii of the pipes are normally distributed about a mean radius a_m. The number N_a of pipes of any radius a can be found from a normal distribution of the appropriate standard deviation. The profile will be built up of narrow pipes having carried dye a short distance, and wider pipes having carried dye a longer distance. The contribution of each pipe to the total dye of the bundle at any section is proportional to the area of each pipe (a^2). For a fixed pressure, the distance the dye travels down a pipe of radius a in unit time is also proportional to a^2 (equation 12.1). We can construct the profile by taking the number N_a of pipes with radius a and multiplying by the contribution of each to the dye content of the bundle (a^2). Then these products are summed cumulatively, starting with the largest a (the front of the profile) and the cumulative totals graphed against the distance travelled (a^2). This has been done in Fig. 13.7 for pipes of mean radius 10 μm, standard deviation 1 μm, and for 2, 5, 10 and 20 units of time. The longer flow progresses, the more the dye in the larger pipes outstrips the dye in the smaller ones, and the flatter becomes the profile. The profile does not closely resemble an error function which is plotted for comparison fitted to the top and bottom values. Nor could an error curve fitted to top and bottom of one profile be likened to that at a later time – the fitting coefficient (translocation coefficient) would increase with time and distance. This multiple-radius dispersion is clearly the kind of dispersion to be expected in a strand of xylem, and of phloem too if the flow is of a simple laminar kind. With sieve plates present in the tubes and constituting the main resistance to flow, the dispersion will be the result of the different pore radii rather than the different tube radii. If wide tubes tend to have wide pores, the effect will be strongly enhanced.

2. PRACTICAL MEASUREMENTS OF TRACER PROFILES

In seeking to learn by some such kinetic analysis what processes are happening in the translocation channel a number of limitations must be kept in mind. The size of the area of application of tracer is going to matter. For example if the application covers a large number of veins of a leaf, these will be at different distances from the petiole, and however simple the profile advancing from each vein, the composite profile in the petiole will be confused by the different times of arrival of the component profiles. The application should be kept local and intense, and it will be well to approach profiles developed from whole-lamina applications with suspicion. Measurements must be made of the developing profile inside the

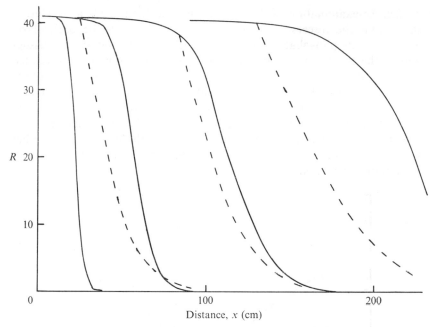

Fig. 13.7. Dispersion of solute profile by flow through a bundle of pipes of different radii. The mean radius is 10 μm and the distribution of radii about this mean is assumed to be normal with a standard deviation of 1 μm. Real bundles of vessels, sieve tubes, etc., will very likely have a larger scatter of radii which will lead to more rapid dispersion. The curves are drawn for 2, 5, 10 and 20 units of time. Error curves are fitted for comparison to the top and bottom values of the profiles, using a constant value of K. The tube-dispersed profiles become less like error curves as they move.

channel, not complicated by exchanges into and out of the channel of the kind allowed for in Horwitz's Models 1, 2 and 3. If such exchange occurs its analysis will tell us about the relations between the translocation machinery and the cells around it, but will obscure the information about the machinery. In designing an experiment to yield kinetic data precautions must be taken to keep these exchanges as small as possible by: (*a*) confining measurements to a short time during the early passage of tracer. The time of application of tracer, whether pulse or continuous, should not affect the shape of the first-arriving front; (*b*) using a tracer that is known to leak as little as possible. The signs to beware of are clear: a straight line on the plot (log R *v.* distance) suggests that leakage may be dominating the kinetics since both models 1 and 3 lead rapidly to such straight-line relations. At the time of Horwitz's paper, and of my early review of this subject (Canny, 1960*b*) such straight lines were believed to be the normal profile. Re-examination of the experiments that led to such a conclusion shows that the tracers used were such as we know to be particularly leaky: ^{32}P in phosphate, ^{42}K, $^{3}H_2O$, rapidly exchanging into tissues outside the translocation channel (Fig. 13.8). It can now be appreciated that the fashion for finding such straight lines arose from the choice of the substances and that a reliable kinetic analysis of profiles in the channel was not to be expected. There were (up to 1960) very few analyses (and most of those not detailed) of the profiles of ^{14}C-sucrose derived from assimilation of labelled carbon dioxide. What there were were either squeezed to fit straight lines, or did yield real straight lines for a particular and important reason.

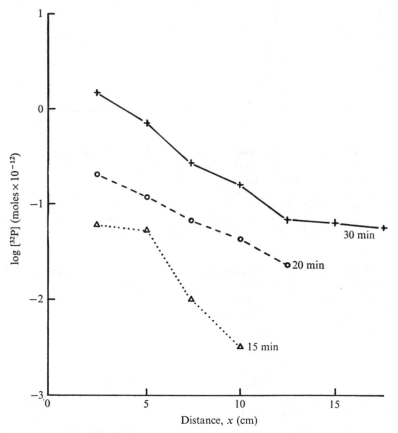

Fig. 13.8. Profiles of ^{32}P-phosphate translocated in the stems of *Phaseolus*. The semi-logarithmic plot of activity v. distance yields roughly a straight line as would be expected for a substance known to leak from the transport channel. Re-drawn from Biddulph and Cory (1957) Fig. 2F.

This reason must be constantly remembered in assessing experimental results of ^{14}C-translocate profiles, and in planning experiments to give such results. *^{14}C-sucrose in transit breaks down to ^{14}CO$_2$ and this can be re-assimilated in green tissues in the light.* From the results discussed in Chapter 12 it is plain that a moderate proportion of the sucrose moved is respired to CO_2 again, and that if the sucrose has ^{14}C in it, so will the CO_2, and this will diffuse widely and rapidly in tissues surrounding the phloem. If these tissues have chloroplasts, as will most of the commonly-used test objects like herbaceous stems and petioles, in the light this ^{14}CO$_2$ will be re-assimilated into tissues around the phloem and when the organ is cut up and extracted this ^{14}C will be assayed along with that in the channel. The effect on the profile will be precisely that of model 1 with a rather high rate of irreversible leakage, and a straight line on the semilog plot will result. This effect and its remedy are beautifully illustrated in the first-published account of ^{14}C-photosynthate profiles, that of Vernon and Aronoff (1952): most of the many profiles in that account were gained from *Soya* bean stems in continuous light. But in one experiment (their Fig. 4) they have data from a plant which was allowed to photosynthesise in the light for 5 minutes and then placed in the dark for a further 15 minutes. The profile is reproduced here as Fig. 13.9. It is compared in their diagram with two other plants kept for the 20 minutes in continuous light of different intensities (Fig. 13.10). They conclude that there is no effect of light on translocation (assessed as speed of the profile) apart from the obvious increase of photosynthesis in going from 80 ft candles to 7000. *But there is an effect on the shape of the curves.* The two lines in continuous light are nearly straight lines; that in darkness is an error function. This is the difference we would expect of re-assimilated ^{14}C turning a pipeline profile into an irreversible-leakage line. In Fig. 13.9 a curve has been fitted * to the data which is the error curve of a diffusion-analogue advance (model 4) whose coefficient is

$$K = 4.6 \times 10^{-2} \text{ cm}^2 \text{ sec}^{-1}.$$

It will be seen that the experimental points approximate very closely to such a curve.

We have reached as it were the watershed of the range of this argument. Behind us lies the jungle of fact which I have tried to order. Before us lies a reasonable country in which these facts will be seen to have their regular places. The comparison of Vernon and Aronoff's two experiments and the value of K coming from the dark set are the first view forward over the pass. Remember and carefully weigh the following:

* This is done in detail in Appendix 1 to serve as an encouragement for others to try it.

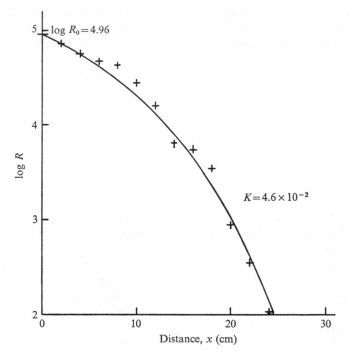

Fig. 13.9. Profile of log [^{14}C-assimilate] $v.$ distance in the stem of *Soya*. Assimilation was for 5 minutes in light of 800 ft-candles and translocation for a further 15 minutes in the dark. An error curve is fitted with $K = 4.6 \times 10^{-2}$ cm^2 sec^{-1}. Re-drawn from Vernon and Aronoff (1952) Fig. 4.

(1) The one careful and complete analysis of mass transfer and its relation to gradient produced the idea that mass transfer behaved like a diffusion analogue.

(2) This data yielded a value of K of 7×10^{-2} cm^2 sec^{-1}.

(3) The first set of clean and detailed data for a non-leaky, translocated tracer carbohydrate measured under conditions precluding extra-phloem re-assimilation yielded a profile suggestive of a diffusion analogue.

(4) The value of K derived from this data was $K = 4.5 \times 10^{-2}$ cm^2 sec^{-1}.

(5) (3) and (4) are absolutely independent of (1) and (2) in concept and in practice.

(6) The agreement of (2) and (4) is not likely to be a coincidence.

Let us now examine the published data of tracer profiles with a wary eye, remembering the conditions that may be expected to produce useful data and those likely to confuse it. But first a word about the plotting of experimental results.

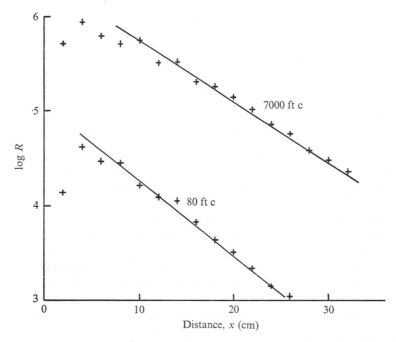

Fig. 13.10. Profiles comparable to that of Fig. 12.9 but with the plants kept in continuous light of 80 and 7000 ft candles during the whole 20 min of translocation. The points are satisfactorily fitted by straight lines. Re-drawn from Vernon and Aronoff (1952) Fig. 4.

Under the influence of Vernon and Aronoff, re-inforced by such examples as those of Biddulph and Cory, measurements of the variation of tracer concentration with distance were commonly plotted on the semi-logarithmic plot. The stimulus for this was the search for the straight-line relation. But whatever the relation, it is still a sensible plot with such data. Measurements of radioactivity are possible over a very wide range of values, and the height of the profile is very commonly three, sometimes as many as six orders of magnitude from peak to background over a distance of a few centimetres. Such values are not accommodated in a linear plot. Compare Fig. 13.3 where values less than 10^3 become lost in the abscissa with Fig. 13.4 which displays the values down to 10, and could easily be extended another several decades. The whole range of measurements can be assessed for the way it follows a predicted curve. Of course the logarithms of the numbers should be plotted on a linear scale, not the numbers on a logarithmic scale, so that the values can be read from the graph with ease. Where other plots have been used, I have replotted them as $\log R$ v. x

from the published curves, or where this was not possible the authors have kindly supplied the primary data.

3. PROFILES NOT CONSIDERED, AND WHY

There are several tracer profiles published in detail which are not relevant to this consideration of kinetics. Nelson and Gorham (1957a & b and 1959a & b) give analyses of the variation with distance of concentration of labelled sugars and amino acids through *Soya* stems. The compounds were introduced through a cut petiole stump and so entered the xylem as well as the cortex and had immediate access only to damaged phloem. With such application there is always the possibility that the tracer will spread in the xylem along gradients of water potential, confusing the phloem profile. Indeed in the 1957 paper glucose and fructose passed a steamed ring – sure evidence that they were not phloem-translocated. The same workers (Nelson, Perkins and Gorham, 1959) have published data on profile shapes of ^{14}C-assimilate in *Soya,* some of which again passed a steamed ring. Those of Clauss, Mortimer and Gorham (1964) are of a similar kind but mostly after times of translocation too long to show the early passage of the tracer wave. The short-time ones are not clear enough for analysis. It is regrettable that the much-used Soya has hollow stems which would allow especially easy gas diffusion and re-assimilation of the respired $^{14}CO_2$. Many of the profiles of Gage and Aronoff (1960) were also produced by feeding to cut petioles, but their labelled-photosynthate data are given below.

Spanner and Prebble (1962) made a detailed study of profiles of ^{137}Cs developing in the petiole of *Nymphoides peltatum*. They used an exterior scanner and were able to measure successive profiles on the same exposed vascular bundle and eliminate variation between different organs. But the times at which the profiles are measured are all very long, from 2 through 36 hours. Leakage dominates the behaviour, as might be expected for such a tracer and for such times. The curves on the semilog plot are linear to slightly concave upwards, a shape which implies the channel is getting larger or the leakage more extensive or the speed greater as the tracer moves.

Straight-line profiles

These have been mostly dealt with above. Nearly all the profiles in Vernon and Aronoff's account are of this kind, except the one in the dark.

Biddulph and Cory's (1957) analysis of a ^{32}P profile has been given (Fig. 13.8); Their similar analysis of ^{14}C-photosynthate is reproduced as Fig. 13.11.

Straight-line profiles are still to be found in many published accounts of

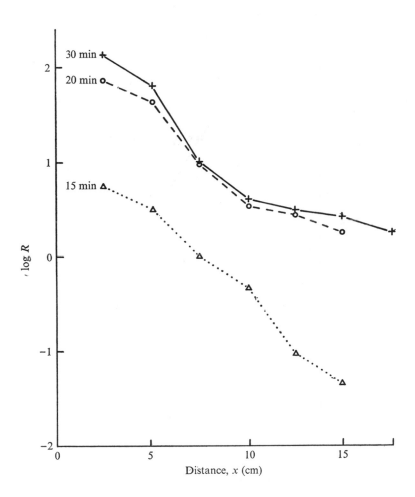

Fig. 13.11. Profiles of log [^{14}C-assimilate] $v.$ distance in the stems of *Phaseolus*. Redawn from Biddulph and Cory (1957) Fig. 2E.

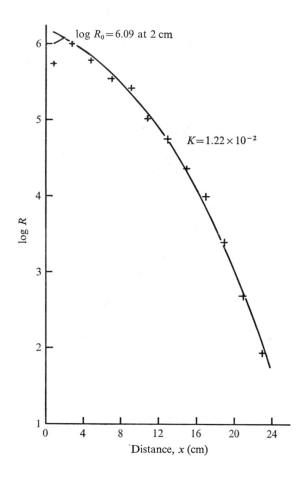

Fig. 13.12. Profile of log [^{14}C-assimilate] in *Soya* after 20 minutes. Re-drawn from Gage and Aronoff (1960) Fig. 1.

^{14}C-assimilate movement, mixed in with profiles of other shapes. The first three figures of Lawton (1967) are practically straight lines, and based on much more frequent sampling of the organs than was usual in early work. Where found after an hour or so such profiles point to leakage.

Error curve profiles

There are now a good many published profiles that can be fitted very closely by error curves. Some of these have been collected as Figs 13.12 to 13.19. Error curves have been fitted and the requisite value of K is marked on each.

Both the data of Figs. 13.14 to 13.17 of Mortimer on *Beta*, and those of Fig. 13.19 by Lawton in *Dioscorea* contain progressive determinations

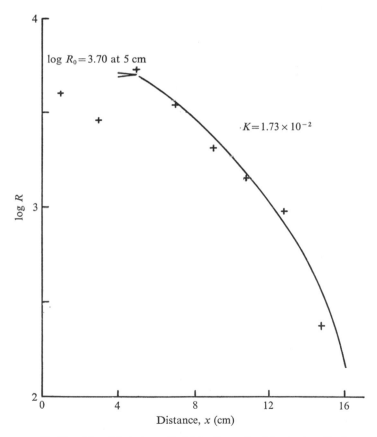

Fig. 13.13. Profile of log [T-photosynthate] in *Soya* after 15 minutes. Re-drawn from Gage and Aronoff (1960) Fig. 8.

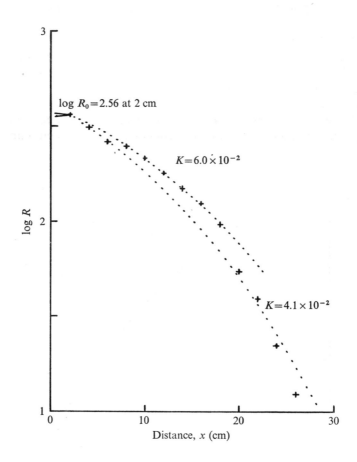

Figs. 13.14 and 13.15. Profile of log [^{14}C-assimilate] in *Beta* petiole after 30 min. Redrawn from Mortimer (1965) Fig. 2, upper points (13.14) and lower points (13.15). Two values of *K* are possible fits, or the profiles may be of the transition type (see Chapter 16).

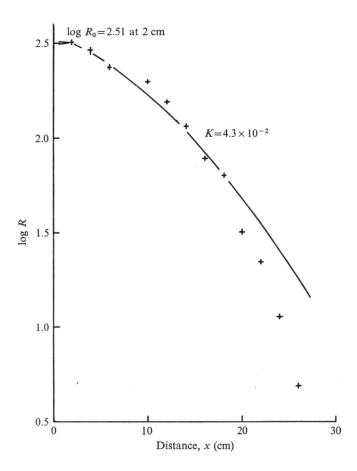

Fig. 13.15. For legend see Fig. 13.14.

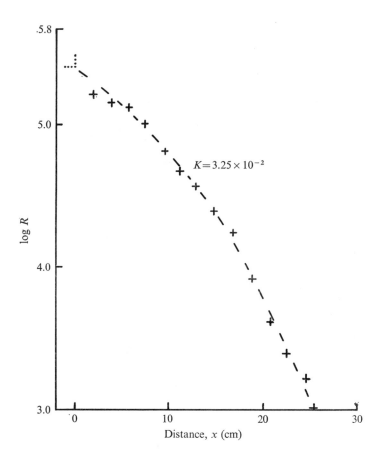

Fig. 13.16. Data of Mortimer (1965) Fig. 5A. Profile in *Beta* petiole at 20 minutes (his circled points). $K = 3.25 \times 10^{-2}$. Re-drawn.

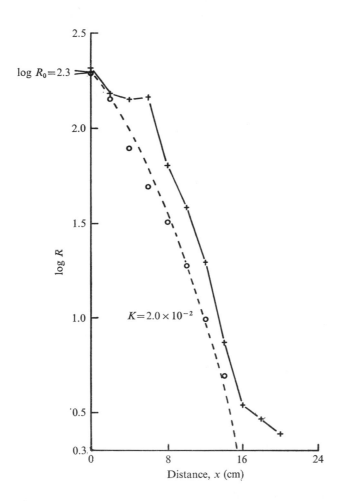

Fig. 13.17. Data of Mortimer (1965) Fig. 2, upper and lower points at 15 minutes. $K = 2 \times 10^{-2}$. Re-drawn.

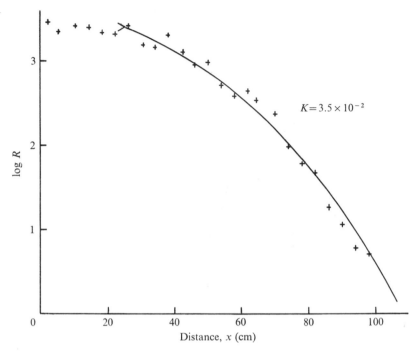

Fig. 13.18. Data of Lawton (1967) for *Dioscorea* ^{14}C-assimilate in the stem after 130 minutes. $K = 3.5 \times 10^{-2}$. Re-drawn.

of profiles developing with time, the former at 15, 20 and 30 minutes, the latter at 4, 5 and 6 hours. The necessity of sacrificing the plant to find the distribution introduces much unwanted variation. Indeed Mortimer draws a graph to show the large differences between plants at the same time, so the agreement or lack of it in values of K during these progressions is not very significant. The most convincing set of error curves are those which can be fitted to the successive profiles of Fisher (1970) for ^{14}C-assimilate in Soya. Here the fits are not only startlingly close, but the same value of K applies from 11 minutes to 35 minutes (Fig. 13.20). The important measurement of the changing profile with time in a single plant was achieved by Moorby, Ebert and Evans (1963) using the more energetic but very short-lived isotope ^{11}C, and continuously monitoring its progress.

Their Fig. 8 shows a series of curves which can be fitted with error curves at the top, but which in the last four centimetres of the profile, curve more quickly downward. I publish here, with Dr Moorby's permission, another set of measurements from this series as Fig. 13.21. Again

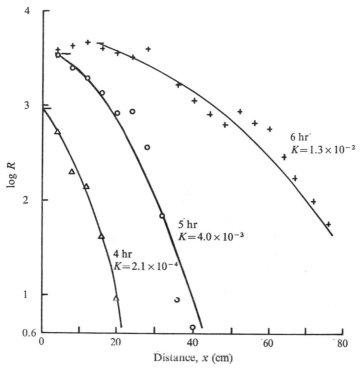

Fig. 13.19. Data of Lawton (1967) for *Dioscorea* ¹⁴C-assimilate in the stem after 4, 5 and 6 hours. Re-drawn.

the profiles drop very steeply at the extreme front. Error curves may be fitted to the rest of the points, and are shown dotted in Fig. 13.21. The fitting constants for the various curves (translocation coefficients) are not the same but increase slightly with time from 3.5×10^{-3} at 17 minutes to 6.9×10^{-3} at 32 minutes. This drift, which is apparent in the data of Fig. 13.19, is even more rapid in Moorby *et al.* (1963), Fig. 8. When there is a drift of the required coefficient, it is to higher values at longer times. A further selection of my own profiles on other plants is given in Figs. 13.22 to 13.24. In these, the precaution has always been taken of darkening the translocation pathway with aluminium foil to prevent re-fixation of respired CO_2. It will be seen that the values of K fall within quite a narrow range, mostly below Mason and Maskell's estimate.

Complex profiles

There remain a good many curves of $\log R$ *v.* x which are neither straight lines nor clear error functions. One was given as Fig. 8.9 in discussing

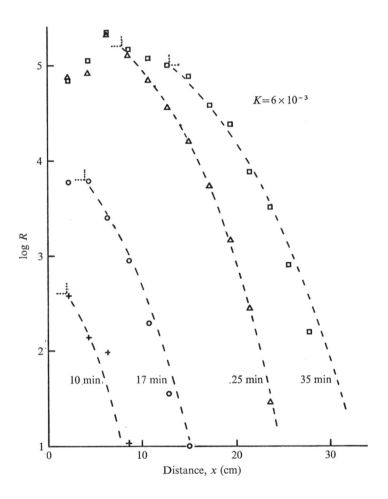

Fig. 13.20. Data of Fisher (1970) for *Soya* stem, ^{14}C-assimilate at successive times. The same fitting constant ($K = 6 \times 10^{-3}$) has been used for all the curves; only the time and maximum radioactivity have varied as shown.

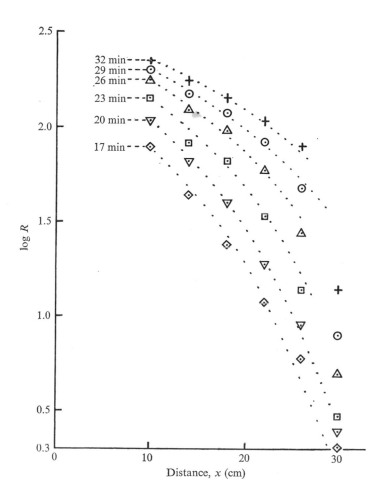

Fig. 13.21. Unpublished profiles of ^{11}C-assimilate developing with time in a single plant of *Soya*, like those of Moorby *et al.* (1963) Fig. 8. The data were kindly supplied by Dr Moorby and are reproduced by his permission. The values of K needed to fitted the dotted curves are:

17 min: 3.5×10^{-3}	26 min: 4.5×10^{-3}
20 min: 3.3×10^{-3}	29 min: 5.8×10^{-3}
23 min: 3.8×10^{-3}	32 min: 6.9×10^{-3}

Each curve drops steeply at the forward end.

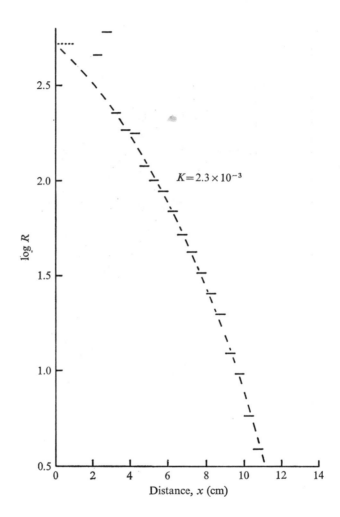

Fig. 13.22. Original data for *Gossypium* petiole log[^{14}C-assimilate] after 60 min trans-location from a spot on the leaf at $x=0$. Temperature 17.5 °C. The first centimetre of the petiole is a swollen pulvinus, which may account for the displacement of R there above the line on which the rest lie. $K = 2.3 \times 10^{-3}$ cm^2 sec^{-1}.

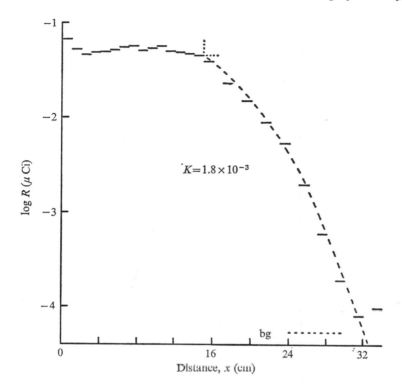

Fig. 13.23. Original data for *Helianthus* stem. $^{14}CO_2$ has been assimilated into a spot on a leaf. The profile is shown in the stem downwards from the application node after $2\frac{1}{4}$ hours. The first 5 cm of the stem contains a steady level of ^{14}C-translocate. Beyond this the front curves steeply down and may be fitted with the error function shown, for which $K = 1.8 \times 10^{-3}$ cm^2 sec^{-1}. The cotyledonary node is at $x = 27$ cm; $x = 1$ to 26 is a single internode.

aphid exudates. Examples of others are given as Figs. 13.25 to 13.28. These have common features and tendencies as follows:

(1) There is often a very small but detectable amount of radioactivity in advance of the main profile. It is in amount less than 1/10,000 of the main peak.

(2) The foot of the main profile is commonly steep, more steep at short times, less steep after long translocation.

(3) With progress through the plant the steep forward front rounds off at the top, merging more gradually into the plateau.

(4) At the top of the rise there is often but not always a small sharp peak.

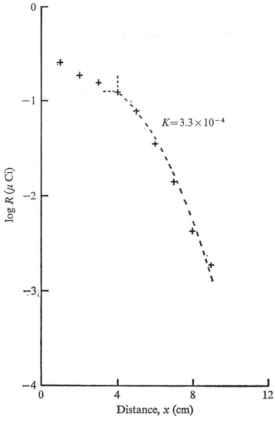

Fig. 13.24a.

(5) Behind the advancing profile there forms a plateau of fairly constant radioactive concentration.

(6) This kind of profile is generally characteristic of fairly slow-moving tracer, and seems to merge at long times, or more rapidly with fast-moving tracer, into the error-curve by the enlargement of the curve of (3) at the expense of the steep foot of (2).

Attenuation with distance

Some advancing profiles probably get lower the further they go (cf. Fig. 13.6), as evidenced by the scanning of Moorby *et al.* (their Fig. 5) where at further distances, the probe records the maximum radioactivity rising

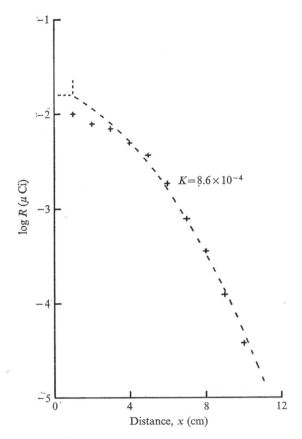

Fig. 13.24*b*. Original data for log[^{14}C-assimilate] profiles in the petioles of strawberry leaves: (*a*) after 70 min, $K = 3.3 \times 10^{-4}$; (*b*) another plant, after 90 min, $K = 8.6 \times 10^{-4}$.

to lower steady values, but this is not a constant feature (their Fig. 7). Nor do profiles of complex type seem to show any attenuation, leaving behind them as they do an apparently saturated plateau.

All these profiles of radioactivity *v.* distance are composite ones formed by the combined contributions of many sieve tubes. We have seen in model 7 how a real strand of conducting tissue may be expected to modify profoundly the simplest flow profile and the question must be posed whether these composite profiles mean anything at all about the concerted workings of translocation machinery in different tubes, or whether they merely reflect the different transport capacities of many tubes. Very little attention has been given to this important question, but what little

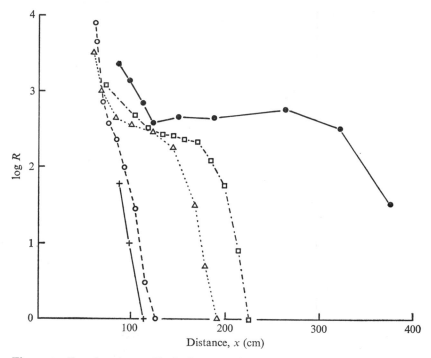

Fig. 13.25. Translocation profiles in the stems of sugar cane at successive times downward from the fed leaf. R is the specific activity of powdered plant material. 1 hour, +——+; 1.5 hours, ○——○; 3 hours, △····△; 4 hours, □····□; 5 hours-●——●. Re-drawn from Hartt *et al.* (1963) Fig. 8A.

evidence there is encourages a faith that the machinery works independently of size. The one set of profiles gained from single sieve tubes via the exuding aphid stylets (Canny, 1961) showed that the radioactivity in a band of sieve tubes in the bark was precisely in step with the exudate from a single tube; that the peak of activity of the time profile was reached in the single tube at the time when the peak of activity of the distance profile coincided with the site of the stylet. Further, none of the published profiles show the kind of progression of shapes that Fig. 13.7 predicts. They are all less dispersed, some of them very much less. More information is needed on this question.

These are the patterns of profile of tracer as it is spread by the translocation process. Knowing what shapes they take it is possible now to consider the question of speed; whether it can be measured, or whether indeed it has any meaning.

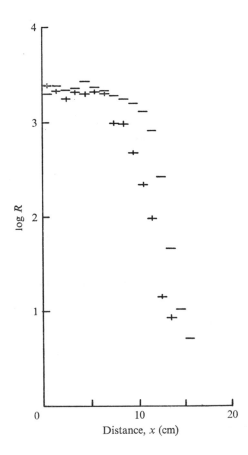

Fig. 13.26. Profile of ethanol-soluble ^{14}C-assimilate in the stem of *Helianthus* down from the fed leaf after 30 min. The plant yielding the profile marked with crosses was boron-deficient. Re-drawn from Lee *et al.* (1966) Fig. 1*a*.

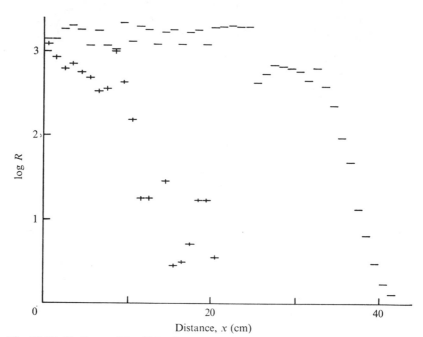

Fig. 13.27. Similar to Fig. 13.26, but at 60 min. Re-drawn from Lee *et al.* (1966) Fig. 1*d*.

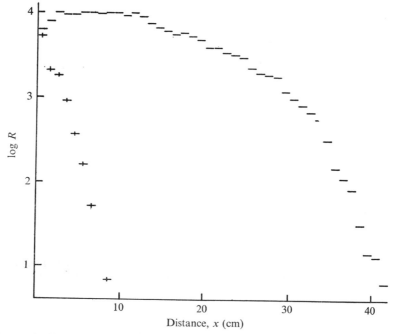

Fig. 13.28. Similar to Fig. 13.26. 60 min. Re-drawn from Lee *et al.* (1966) Fig. 1*e*.

14. The speed of translocation

Says Tweed to Till –
'What gars ye rin sae still?'
Says Till to Tweed –
'Though ye rin with speed
And I rin slaw,
For ae man that ye droon
I droon twa.'   Anon.

From the outset it has been emphasised that translocation is a process of the transfer of dry weight. In most such transfer processes the question of speed is held to be of little importance. In getting water from a pipe one is interested in the quantity that comes out in a given time, not in the fact that the speed in the centre of the pipe is twice the mean speed; the physician considering the flow of blood concentrates his attention on the rate of pumping, the pressure the pump develops and its rate of working, not the speed of flow; the electrical engineer considers the quantity of current (amps) a wire carries, not the speeds of the electrons. Even in the flow of xylem water the plant physiologist does not use speeds as a useful measure but assesses the quantity taken up by the roots or lost by the leaves. These are relatively simple systems where the relation of speed to rate of transfer is clearly understood. But once that understanding was established by confirming that the speeds agreed with the assumed models, the speeds were not used to measure the processes. In the study of translocation, however, the question of speed has a fascination that defies logic. Here there is no agreement about a model of working nor the relation that the transfer rate may have with a speed. There is therefore the legitimate area of exploration to try to find such a relation. A simple example is the flow relation of equation 1.1. But alongside this endeavour is the feeling that one ought to be able to *measure* translocation by its speed. It may prove so but the relation must be established first, and the usefulness of the measure demonstrated.

There are systems in which the rate of transfer has no simple relation with speed. One is that of diffusion, with which translocation has been seen to have many similarities. If there is any speed associated with diffusion, it is the speed of the molecules and depends only on temperature. But at a fixed temperature and molecular speed it is possible to get any rate of transfer of mass from zero (with zero gradient) to nearly infinity (over a very short distance with a steep gradient). If the analogies of translocation with diffusion extend as far as they seem to, both in the steady-state relation of mass transfer and gradient, and in the non-steady-state advance of tracer molecules, we should be prepared for a similar lack of connection between speed and transfer. Tracers may move rapidly but there may be no transfer of mass; there may be rapid transfer of mass with tracers moving at the same speed as before. This was explained qualitatively in Fig. 5.3, and from the quantitative analysis of the tracer advance in the last chapter there is added reason to expect this kind of speed independence.

Since I last reviewed the problem of speed (Canny, 1960b) much more information has become available which is relevant to the problem, but there still are basically only three ways of doing it: by the progress of a wave of chemical concentration; by the physical detection of a moving stream; and by the progress of tracer substances through the system. The last is the most popular and embraces many diverse methods.

I. CHEMICAL MEASUREMENTS OF SPEED

The one comprehensive attempt to follow a wave of sugar from the leaves of a tree through the phloem of the trunk after a day's photosynthesis (that of Hüber *et al.*, 1937) has already been treated at some length in Chapter 8. Hüber *et al.* drew a straight line on Fig. 8.6 connecting the positions of the troughs of concentration on the first afternoon, and used the slope of this line to give a measure of the speed of advance of the trough: 11 m in 4 hours, or 3.6 m hr^{-1}. The measure of speed may not be a valid one. Reasons have been given above for believing that the results do not show only the simple effect of the advancing wave of fresh photosynthate. Other reasons have been advanced in another paper (Canny, Nairn and Harvey, 1968), notably the very high rate of specific mass transfer implied by the passage of an 18 per cent sugar solution at a speed of 3.6 m hr^{-1}. Zimmermann in correspondence has questioned our assumption of the phloem area of the tree trunk, and he is probably right. Reducing the assumed area of phloem to that given by Münch (1930) for a large tree of *Quercus rubra* (Table 1.1), still leaves the rate of sugar movement very large indeed: the similar sets of data recorded by Ziegler (Fig. 3.1) and Zimmermann appear to yield similar high speeds but are less convincing. It seems that so far chemical measurements of this kind have yielded only doubtfully useful results.

An interesting refinement of this method has been devised by Zimmermann (1969), using not the sugar concentration, but the relative concentrations of the mixture of sugars translocated down the trunk of *Fraxinus americana*. He observed that the relative concentrations of the sucrose and its related oligosaccharides were not constant in the phloem exudate, but varied slightly with height and time. It seemed that the different sugars were supplied to the phloem in slightly different amounts at different times of day, and that as the sugars travelled down the trunk the concentration ratios travelled like a marker of the wave of sugar. He assumes without proof that this effect is produced by the common speed of the three sugars preserving the differences, not (as Trip *et al.*, 1965, assumed) by the three sugars travelling at different speeds and producing a resultant speed of the ratio by interference. The speed of the ratio wave varied from 30 to 70 cm hr^{-1}.

2. PHYSICAL MEASUREMENTS OF SPEED

There is now one published account claiming the physical detection of a flowing stream in the phloem and offering an indirect estimate of its speed. This is the elegant heat-transfer experiment of Ziegler and Vieweg (1961). Hüber successfully used a heat pulse to measure flow in the xylem, detecting changing temperatures at thermal junctions up and down stream from a warmed piece of exposed xylem. Ziegler and Vieweg applied a similar method on a small scale to the phloem strand dissected out of a *Heracleum* petiole bundle, still attached at both ends. Two thermal junctions were in contact with the lower side of the phloem strand 0.5 cm apart and the difference in temperature of the two junctions was displayed on a galvanometer. Heat was fed into the bundle by focusing a small spot of light from a microscope lamp. The asymmetry of the geometry and contacts was such that they could not use the direct method of finding the point between the junctions where the heat transfer by combined conduction and flow gave the same temperature at each junction. Rather they found a null point of the galvanometer which was the mean of the readings when first one junction position was heated, and then the other. The light spot was then moved to that position between the junctions at which the galvanometer indicated this null point. Now the flow of heat by whatever pathway(s) to the two junctions was the same, and interrupting a hypothetical flow of solution within the bundle should change the temperatures. They found that on cutting the petiole strand on the leaf side of the assembly, the junction on the plant side always got cooler by about 0.125–0.250 °C within a few seconds. This is what would be expected if the downstream junction owed its temperature partly to the arrival of heat transported towards it in a flowing stream from the leaf blade. They did not unfortunately check that it was not purely an effect of a surge produced by cutting, which might have been done by showing that a cut on the plant side of the assembly also produced a cooling of the same junction. If on the other hand the junction on the leaf side now got cooler, this would suggest the effect was due rather to a surge towards the cut.

They fix an order of magnitude for the effect by simulating the system with a strand of hemp with water flowing over it, and finding what speed of flow of the water produced similar temperature changes. This speed was about 35–70 cm hr^{-1}.

3. TRACER METHODS OF MEASURING SPEED

Table 14.1 collects the available data on speed measurement by a variety of methods. Not all will be discussed in detail, but the differences in method need to be underlined, and some of the more interesting points will be mentioned. The Table follows an approximate order of increasing refinement of measurement, from the early use of the furthest distance at which trace-substance could be detected, through various destructive-sampling methods of following profiles, to the non-destructive scanning of living plants. The last offers the most hope of a precise result since it reduces variation between plants, but the destructive sampling methods have been made to yield precise data from many replicates of uniform plants.

It is obviously of prime importance to all such determinations to know: (1) the shape of the profile with distance; (2) whether the shape changes with time; (3) whether the height of the profile gets less as it moves. Even with simple flow of coloured solution through a tube these three things must be known in order to follow the speed of flow, as Fig. 13.6 shows. Early work on speed of tracer movement did not take these into account and the results are of limited value (Table 14.1, Part III 1a).

The paper in which Nelson, Perkins and Gorham (1958) show a very far advance of ^{14}C assimilate in the first 30 seconds, calculating a speed of up to 2400 cm hr^{-1}, has been often quoted as a firm figure for translocation speed. However, they make it plain that the main 'slow front' of labelled assimilate lagged far behind; that the fast component was not continuous through the tissue but seemed to skip patches; and, in their 1959 paper, show that this fast component passes a steamed ring and is probably moving in the xylem.

The only attempt to obtain a speed measure for a tree by these tracer means for comparison with Hüber's chemical-wave measure is that of Canny *et al.* (1968). Physical difficulties of application, harvest and assay, and the availability of large numbers of unwanted trees, are limitations that prevented this test being made earlier. It is regrettable that the species is so different from *Quercus rubra*, though one similar trial was done with another *Quercus* species and gave the same answer: that ^{14}C-assimilate moves much slower in tree phloem than the Hüber figure of 3.6 m hr^{-1}. The view is advanced that a tree is a slow-paced organism operating at a

TABLE 14.1. *Measurements of speed of translocation*

Author	Plant	Substance	Method	Result – speed cm hr^{-1}
I. Chemical measurement				
Hüber, Schmidt & Jahnel (1937)	*Quercus rubra*	Sugar	Wave of photosynthate (See Fig. 8.2)	360
Zimmermann (1969)	*Fraxinus americena*	Sucrose, stachyose raffinose	Progress of ratio of sugar concentration	30–70
II. Physical measurement				
Ziegler & Vieweg (1961)	*Heracleum mantegazzianum*	Transloc. stream?	Heat input and balanced thermal junctions	35–70
III. Tracer methods				
III.1. Destructive sampling				
III.1a. Distance reached by furthest detectable tracer				
Schumacher (1933)	*Pelargonium*	Fluorescein	(See Fig. 13.8)	up to 35
Biddulph & Cory (1957)	*Phaseolus*	^3H_2O		87
		$^{32}PO_4$		87
		^{14}C-assimilate		107
Nelson & Gorham (1957b)	*Soya*	^{14}C-sucrose		130–300
Nelson, Perkins & Gorham (1958)	*Soya*	^{14}C-assimilate	Fast front	*1900–2400
			Slow front	14–26
Bowmer (1960)	*Gossypium*	Fluorescein	(See Fig. 7.2)	2–6
Webb & Gorham (1964)	*Cucurbita*	^{14}C-assimilate		290

TABLE 14.1. (Contd.)

Author	Plant	Substance	Method	Result – speed cm hr^{-1}
Webb & Burley (1964)	Acer negundo seedling	Stachyose		
		^{14}C-assimilate		55
	Fraxinus americana seedling	Stachyose		21–23
		^{14}C-assimilate		
	Verbascum thapsus	Stachyose		17
		^{14}C-assimilate		
	Cucurbita pepo	Stachyose		28
Thrower (1965)	Soya	^{14}C-assimilate	Released from a cooling block	17 $\pm$ 4
Parker (1965)	Macrocystis	^{14}C-assimilate		65–78
		Fluorescein		1.0
Canny et al. (1968)	Acer pseudoplatanus	^{14}C-assimilate		av. 2 (max. 17)
Geiger et al. (1969)	Beta vulgaris	^{14}C-assimilate	Fastest speed obtained as specific activity increased	54
III.1b. Progress of profiles				
Vernon & Aronoff (1952)	Soya	^{14}C-assimilate	Log straight-line type	84
Canny (1961)	Salix	^{14}C-assimilate	Exudate; complex profiles. Comparison of slopes of $\log R \, v.x$ and $\log R \, v.t$	2–14
Canny (1962a)	Salix	^{14}C-assimilate	$^{14}CO_2$ output; complex profiles, time and distance	0.3–1.6

Hartt et al. (1963)	Saccharum	^{14}C-assimilate	Complex profiles	42–150
Whittle (1964a)	Pteridium	^{14}C-assimilate	Diffusion profile advance	2.5–20
Wardlaw (1965)	Triticum	^{14}C-assimilate	Successive R $v.x$ profiles at short time intervals. Time to reach peak height in each section	87–109 in stem
Mortimer (1965)	Beta	^{14}C-assimilate	Progress of common intercept, said to be linear profiles but see Fig. 13.14, 13.15.	50–135
Lee et al. (1966)	Helianthus	^{14}C-assimilate	Progress of complex profiles	60
Evans & Wardlaw (1966)	Lolium	^{14}C-assimilate	As Wardlaw (1965)	42–105
Wardlaw (1967)	Triticum	^{14}C-assimilate	As Wardlaw (1965)	33–72
King et al. (1968)	Pharbitis	^{14}C-assimilate	As Wardlaw (1965)	33–37
III.2. Non-destructive sampling				
Bachofen & Wanner (1962)	Phaseolus	^{32}P	2 probes, timing of passage of profile (? linear)	Very variable 50–275
Moorby et al. (1963) and Evans et al. (1963)	Soya	^{11}C-assimilate	Scanning of live plant by external probes; progress of mid point of time profile (? Diffusion type)	Mean 130 60.4 ± 7.2
Willenbrink & Kollmann (1965)	Metasequoia	^{14}C-assimilate	2 external probes, time for passage of same R	48–62

* Shown in their 1959 paper to be moving in the xylem, but often quoted

more leisurely and majestic rate that the herbaceous systems mostly studied. As was pointed out in Chapter 3, if the driving potential for translocation is a gradient of sucrose concentration, then this gradient must be very flat in a tall tree and the mass transfer down it may be expected to be slower than in short-length herbaceous systems. It can now be added: even if the speed components are the same.

The results of Part III 1*b* are more reliable, being based on data of

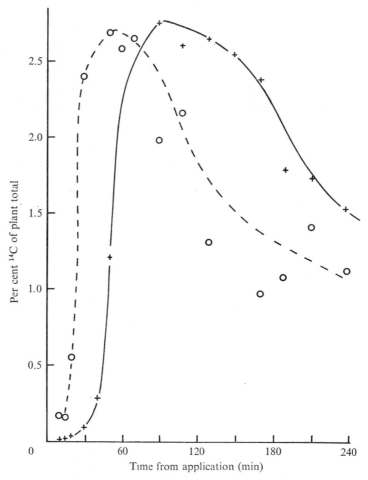

Fig. 14.1. Pattern of accumulation of ^{14}C with time at two points in wheat plants fed with $^{14}CO_2$ on the terminal 19 cm of the flag leaf blades, obtained by harvesting a number of replicate plants at successive times after feeding. The time course of ^{14}C content of the flag-leaf section adjacent to the ligule is shown $\bigcirc$----$\bigcirc$; that of the stem section adjacent to the ear, +——+. Re-drawn from Wardlaw (1965) Fig. 2.

profile advance, sometimes very complete. Those of Canny (1961, 1962a) were an early attempt to assess profile passage past a point by comparing the log R v. time curve at a point on the stem (measured once by aphid-stylet exudate, and later by output of respired $^{14}CO_2$) with the log R v.

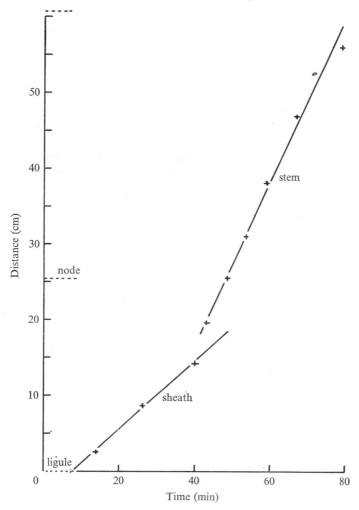

Fig. 14.2. Movement of the ^{14}C profile from the flag leaf blade to the ear. The distance reached by the point of inflection of profiles like those in Fig. 14.1 is plotted against the time from application, giving two lines, one for the leaf sheath and one for the stem, whose slopes are the speeds of advance of the profile through the two regions. The speed in the sheath is 39 cm hr^{-1}; that in the stem, 87 cm hr^{-1}. Re-drawn from Wardlaw (1965) Fig. 4.

distance curve within the phloem after it had passed. The profiles were of the complex type. The method is only valid, as Spanner (1962) points out, if the profile does not change in size and shape as it moves, as these probably did not.

In the hands of Wardlaw (1965) and three subsequent papers, the destructive-sampling method of measuring profile advance has reached its greatest precision. The method is to use a large number of uniform replicates, harvesting a batch every few minutes of translocation and obtaining an average for the tracer content of each section of leaf and stem. A plot of R $v. t$ is built up for each section (Fig. 14.1), and the progress of these curves through the distance of translocation is followed. Wardlaw plots the time of passage of the half-saturation radioactivity against distance, and gets a series of points (Fig. 14.2). The line through them has the speed for its slope.

The non-destructive methods of Part III.2 are not necessarily as accurate as the carefully-sampled destructive ones unless full advantage is taken of the chance to minimise the variation. Moorby *et al.* (1963) report what must be the ideal experiment of this kind. Not only is ^{11}C so energetic that it can easily be monitored from outside the plant at short time intervals, but it has such a short half-life (20.5 min) that it disappears from the plant in a few hours and the same plant may be used again. These data have been given as Fig. 13.21 as an example of profile shape. Evans, Ebert and Moorby (1963) analyse them to give a speed measure, using a plot like Fig. 14.2. They plot the time for the peak of the first differential of activity with respect to time to reach various distances in a short pulse experiment. This gives the speed of that part of the activity travelling with average velocity and average delay. The slope of the line gives the speed as 60.4 ± 7.2 cm hr^{-1} while its intercept gives a measure of the delay in getting assimilated carbon into the sieve tubes, 5.3 ± 2.1 min.

All these measures (even the worst of them) are direct in the sense that they are attempts to measure directly the speed of something which is observed experimentally to move. They should be clearly distinguished from indirect measures that assume some mode of working of the system and relate measured attributes to some speed of a part of the model, such as the concentration-speed relation of equation 1.1. Another analysis of the indirect kind is made by Spanner and Prebble (1962), showing how the model 1-type irreversible-leakage system produces straight lines on the semi-log plot, and how the advance of these straight lines may be related to the assumed velocity of flow of solution in the model. They make the important point that the successive intercepts with a horizontal distance axis do *not* lead to a measure of this velocity. The intercepts must be with a line sloping downward at an angle which is the exponent in the equation relating concentration to distance. However, neither their data nor any-one else's are sufficiently precise to find what this angle is and use it for such a derived velocity.

It can be concluded from this survey that experiments of the three types lead to the revealing of some kind of speed component associated with translocation. It is variable from one plant to another, and even from one part of a plant to another (Fig. 14.2) and so to be useful must be measured in many replicates. This has been achieved in both destructive and non-destructive sampling, though with considerable labour. Determinations of the speed that are not carefully related to a knowledge of the profile shape and attenuation, nor statistically sound, are almost meaningless but will probably fall in the range $1-150$ cm hr^{-1}. What relation, if any, these speeds bear to the transfer of mass has still to be investigated. Whether the speed stays much the same with large or zero transfer of dry weight, as in a diffusion analogue, or whether the speed is directly related to how much sugar is moving, is a crucial question that awaits experimental attack. In order to use this speed in calculations it must be expressed in units con-sistent with the other units employed. From here on speeds will be quoted in centimetres per second, and the reader may conveniently remember that 100 cm hr^{-1} is equal to 2.78×10^{-2} cm sec^{-1} or 4.0×10^{-2} cm sec$^{-1} = 144$ cm hr^{-1}.

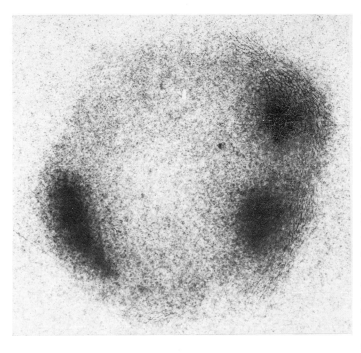

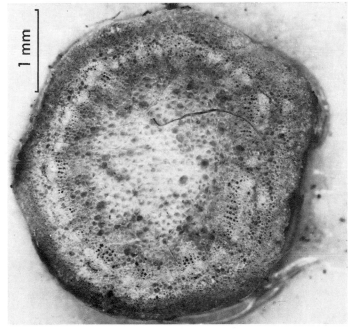

Plate 1. Section of freeze-dried petiole of vine carrying ^{14}C-labelled assimilate from the leaf lamina, and the autoradiograph made by exposing it to X-ray film. This crude procedure can give quite precise localisation of the label at the tissue level.

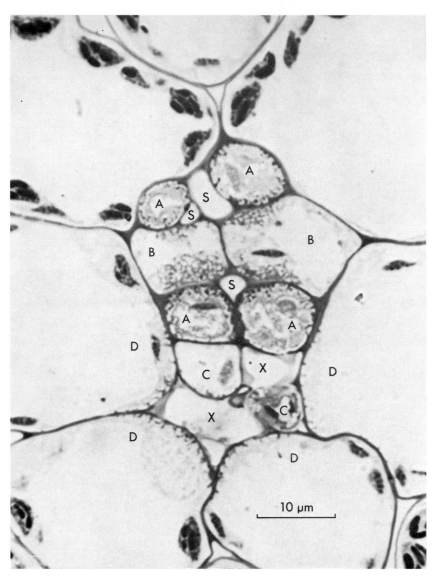

Plate 2. Optical micrograph of a transection of a minor vein of a leaf of *Anacyclus pyrethrum* showing four types of perivascular transfer cell. The vein shows three sieve elements (S) and four associated A-type cells (A). Between these lie two B-type transfer cells (B) with ingrowths proximal to the sieve elements and their neighbouring A-cell walls. Xylem elements are present (X) having associated with them two C-type cells (C), and four D-type cells (D) of the bundle sheath. Reproduced by permission from Pate and Gunning (1969), Fig. 2.

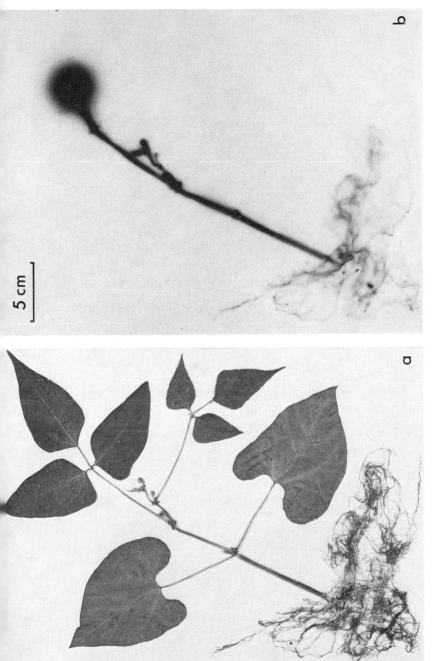

Plate 3. (a) Plant of *Phaseolus* which assimilated $^{14}CO_2$ at a patch on a single lateral leaflet of the first trifoliate leaf, and was allowed to translocate the products for 4 hours before it was oven dried. (b) The autoradiograph formed by exposing the plant to X-ray film for 2 days. Note the absence of labelled translocate from the mature and half-grown leaves, and its presence in the young buds and the roots.

Plate 4. The severed stylet of an aphid that has been feeding on a juniper twig. The mandible-segments of the stylet spring outwards from the cut end of the maxillary segments. The latter, remaining locked close together, enclose the stylet canal, forming a tube through which the sieve tube sap exudes. A drop of sap has collected on the maxillary segments. Reproduced by permission from Fig. 1 of Kollmann and Dörr 1(966).

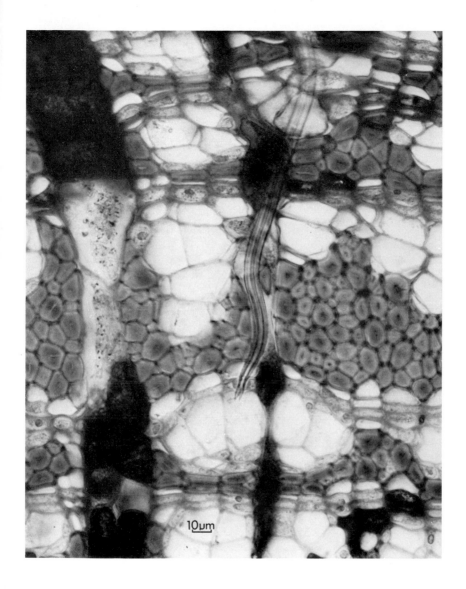

Plate 5. The tips of the maxillary segments of the stylet end in a sieve element. The canal can be seen through which the sap exudes. Copyright 1961 by the American Association for the Advancement of Science. Reproduced by permission from Zimmermann (1961) Fig. 4.

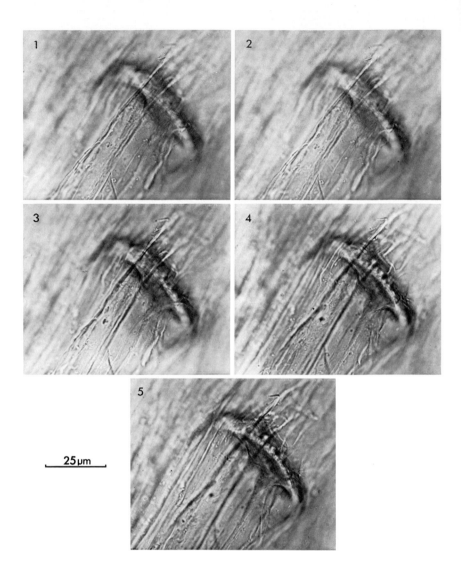

Plate 6. Series of bright-field photomicrographs focusing through a sieve plate of *Cucurbita pepo* after fixation in 6 per cent glutaraldehyde, dehydration, and mounting in araldite. Some fibrils appear to pass through the sieve plate. The successive planes of focus are 2 μm apart. Reproduced by permission from Thaine, Probine and Dyer (1967) Plate 6, Figs. 17–21.

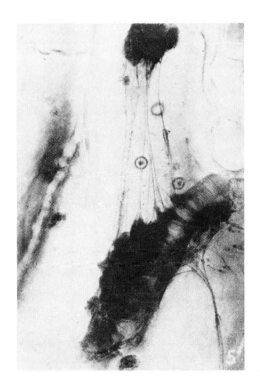

Plate 7. The fibrous protoplasm visible in the damaged mature sieve element of *Fraxinus americana* (× 700). A lump of slime is seen at the top and plastids among the threads. Reproduced by permission from Crafts (1939*b*) Fig. 5.

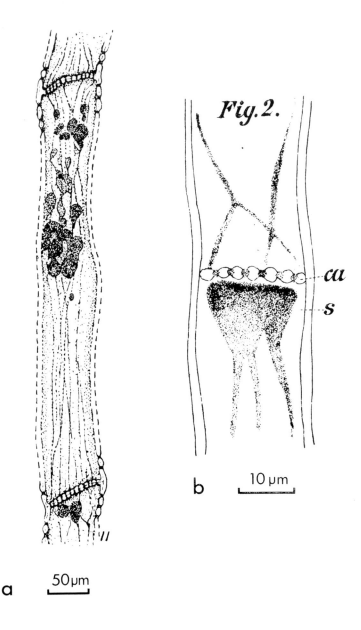

Plate 8. Drawings of the fibrous state of the protoplasm in damaged sieve elements. (*a*) In cucumber, fixed with hot water followed by 50 per cent alcohol and stained with water blue. From Crafts (1932) Fig. 11. (*b*) in *Ecballium elaterium* fixed with alcohol. From Fischer (1886) Fig. 2. Reproduced by permission.

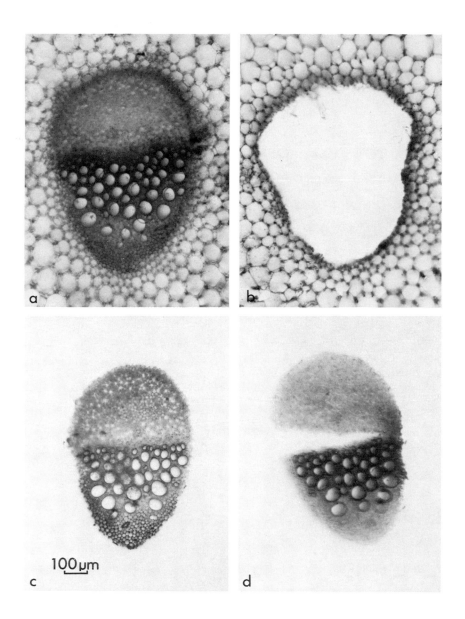

Plate 9. The easy separation of the vascular bundle out of the petiole of *Heracleum mantegazzianum* and its division into xylem and phloem portions. (*a*) Bundle in cortex; (*b*) ground tissue from which the bundle has been removed; (*c*) isolated bundle; (*d*) divided bundle. After Ziegler (1958) Fig. 1.

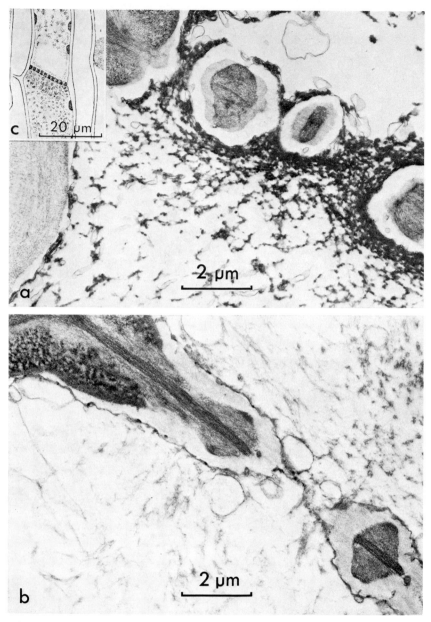

Plate 10. Fixation images of sieve tube contents obtained by boiling and freezing *Cucurbita* stems. (*a*) and (*c*) show the electron image and Fischer's (1886) drawing of the light image formed when boiling water is used to stabilise the sieve-tube contents. (*a*) was post-fixed in acrolein and osmic acid. The plasma filaments are seen to be of two kinds: larger, more diffuse, tubular elements; and smaller, denser strands. The contents have clearly contracted away from the wall at the pore, presumably on immersion in the boiling water. (*b*) was frozen and then thawed in 3 per cent glutaraldehyde and post-fixed in osmic acid. The distinction of the filaments is less clear, and there has been no contraction. Fischer's choice of boiling as a means of stabilising the contents seems fully justified. (*a*) and (*b*) were kindly prepared by Dr T. P. O'Brien. (*c*) is Fig. 40 of the Fischer paper, and is listed by him as 'oblitierende Siebröhre'. It would probably nowadays be called 'mature'.

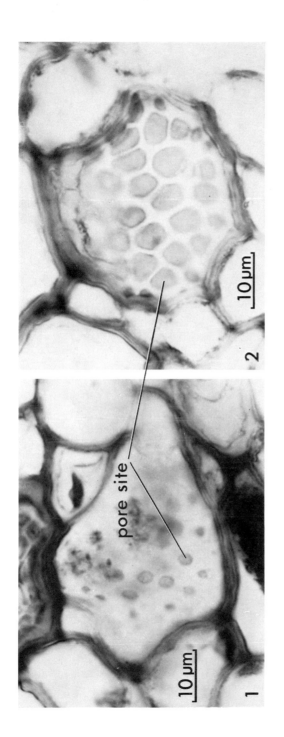

Plate 11. Early stages in the development of sieve pores in the plates of *Robinia* secondary phloem. The callose platelets at the site of the future pore have expanded in 2 from the small beginnings shown in 1. Copyright by the University of Chicago Press. All rights reserved. Reproduced by permission from Esau, Cheadle and Risley (1962) Figs. 1 and 2.

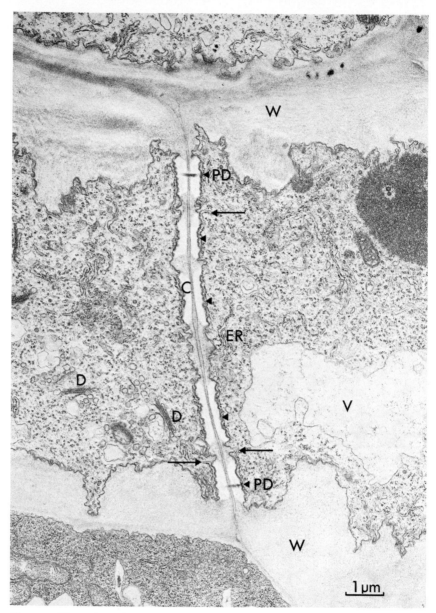

Plate 12. Formation of sieve pores. (*a*) Longitudinal section through the developing sieve plate of *Cucurbita maxima*. Callose platelets (C) occur at the positions of future pores. The endoplasmic reticulum at this stage shows associated ribosomes (ER), and is in close association with the plasmalemma over the callose platelets at the pore sites (small arrowheads). In some places this ER is seen bent away from the uncallosed parts of the young sieve plate (long arrows). W, wall; V, vacuole; PD, plasmodesma; D, dictyosome.

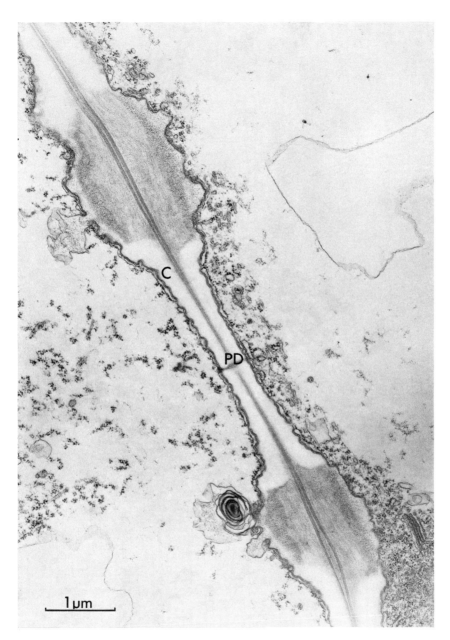

(*b*) View of the pore site and associated cisternae at higher magnification. C, callose platelets; PD, plasmodesma. Reproduced by permission from Cronshaw and Esau (1968) Figs. 1 and 14.

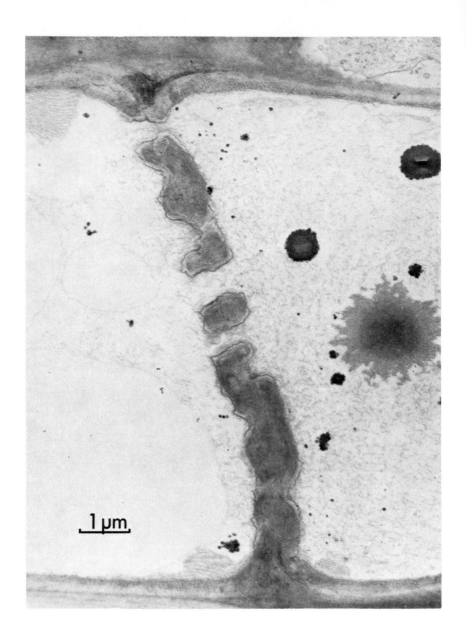

Plate 13. Electron micrograph of sieve plate and the tube contents on either side in *Gossypium hirsutum* fixed with glutaraldehyde–acrolein at 0 °C penetrating the uncut phloem sideways from two longitudinal cuts. Reproduced by permission from Shih and Currier (1969) Fig. 6.

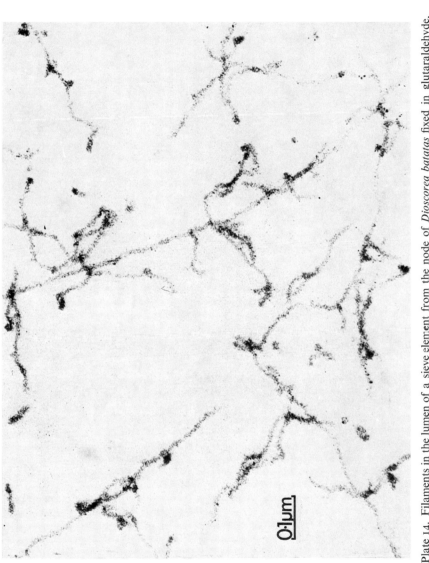

0·1μm

Plate 14. Filaments in the lumen of a sieve element from the node of *Dioscorea batatas* fixed in glutaraldehyde, followed by osmium tetroxide. Reproduced by permission from Behnke and Dörr (1967) Fig. 8.

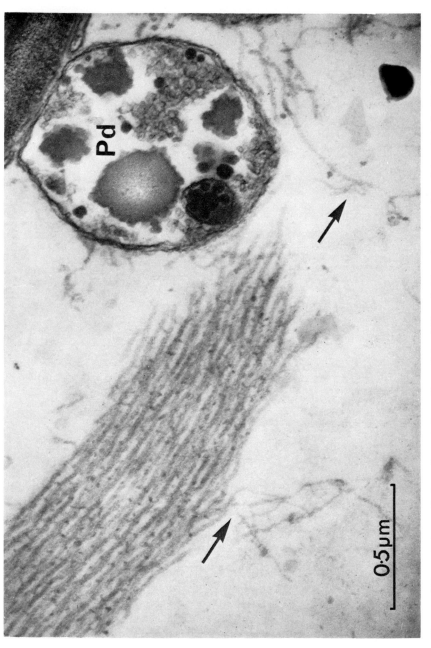

Plate 15. Filaments in small sieve elements from the petiole of *Nymphoides peltatum* fixed in glutaraldehyde. The bundle is composed of filaments of about 240 Å diameter each. They appear to fray out to finer, banded filaments (arrows). Note the unbroken plastid Pd which is evidence of relatively gentle, undamaging fixation. Reproduced by permission from Johnson (1968) Fig. 6.

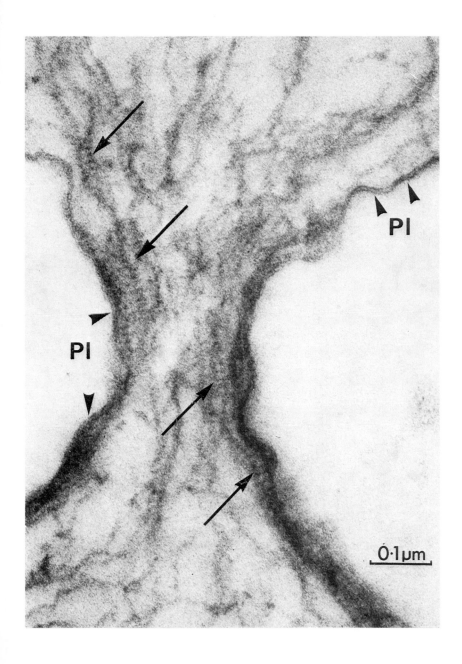

Plate 16. Filaments in sieve pore of *Nymphoides peltatum* again showing banded filaments. Pl, plasmalemma lining pore. Reproduced by permission from Johnson (1968) Fig. 8.

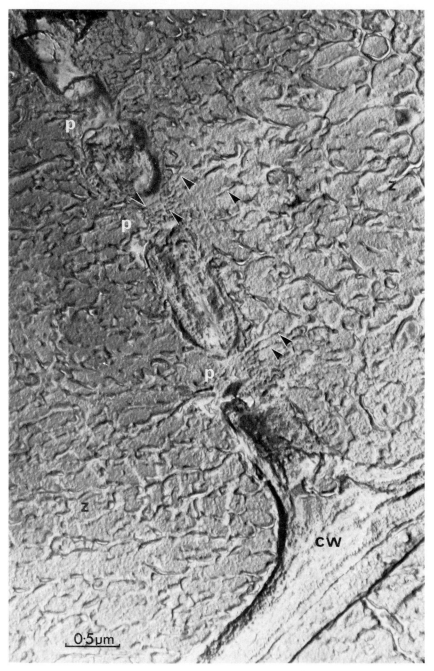

Plate 17. Replica of part of a frozen-etched plate from a sieve tube similar to those shown in Plates 15 and 16. No fixation, just freezing. Banded filaments appear near the arrows. cw, cell wall; p, pore; z, eutectic. Reproduced by permission from Johnson (1968) Fig. 16.

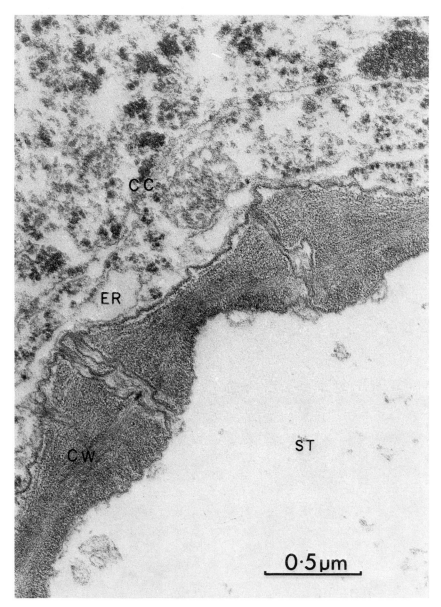

Plate 18. Same material and fixation as in Plate 13, showing branched plasmodesmata on the companion cell side of a boundary between companion cell and sieve tube. The ER system of the companion cell appears connected to the pore. CC, companion cell; ER, endoplasmic reticulum; ST, sieve tube; CW, cell wall. Reproduced by permission from Shih and Currier (1969) Fig. 8.

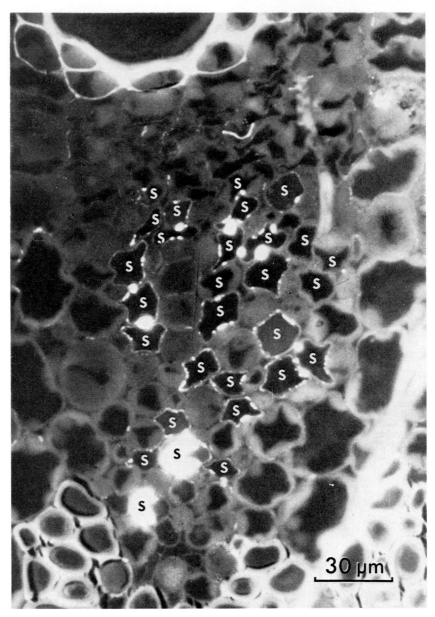

Plate 19. Photograph by UV light of transverse 2-μm section of the secondary phloem of a stem of *Gossypium* stained for callose fluorescence. The cells are marked 's' which at some level in the series of 64 sections showed by the presence of a sieve plate that they were sieve tubes.

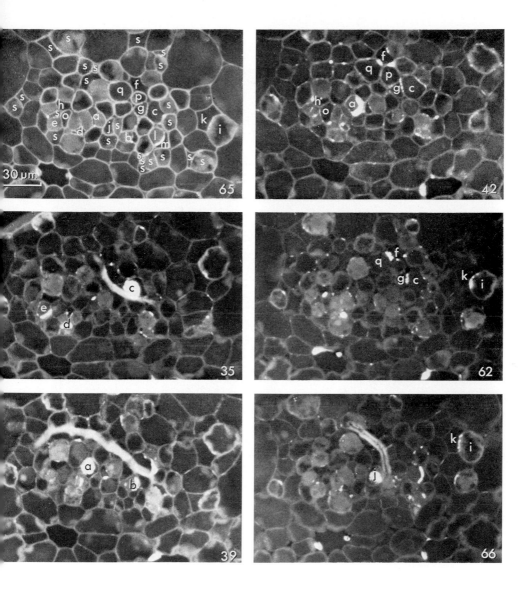

Plate 20. Selected sections from a series through the inner phloem of a bundle of a *Cucurbita ficifolia* petiole. In section 65 those cells are marked with letters which, in some section of the series, are found to be sieve elements. The other sections, 35, 39, 42, 62, 66, show some of this evidence for particular cells, whose letters correspond to those of section 65. Section 65 contains a number of other sieve elements, marked 's', the proof of whose character is elsewhere in the series.

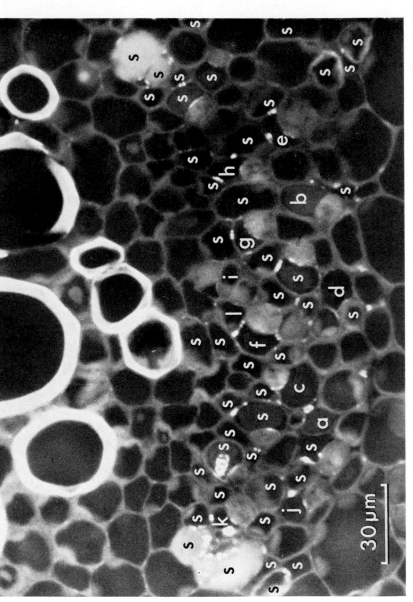

Plate 21. Fluorescent callose-stained resin section of the outer phloem of the same bundle as in Plate 20. Cells marked 's' are sieve elements. Cells marked 'a–l' are identified as sieve elements from their appearance in sections of Plate 22 (3-μm sections).

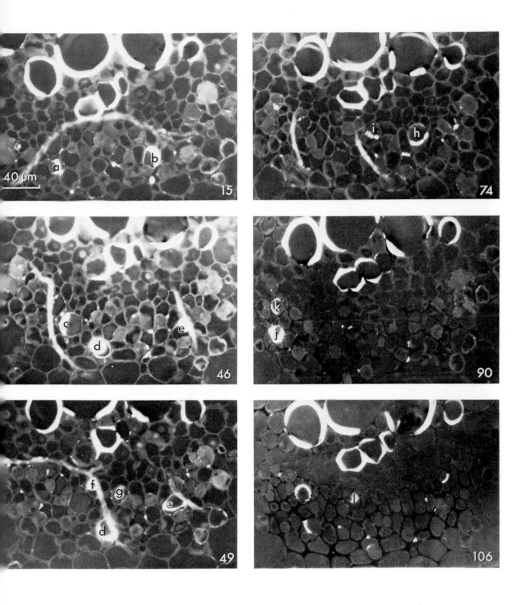

Plate 22. Sections from the series of which Plate 21 is one, providing evidence to identify the cells marked 'a–l' as sieve elements.

Plate 23. (*a*) Fruits of *Kigelia aethiopum* at Paraa on the Nile in July. Note the inflorescence axis at the right which has had no fertile flowers; the nodes are triple and it has withered to a thin string. The fruit in the foreground is formed from a flower near the morphological base (top) of the axis which extends for many nodes beyond the fruit peduncle. The axis above (in space) the fruiting node is thick. The next node above the fruit has two flower scars on either side at the same level and one on the front face of the axis some way below the other pair. Compare the internode lengths on the infertile axes with those on the thick axes bearing fruits. At the top of the picture two fruits have developed at the same level on an axis.

(*b*) The inflorescence axis between a branch and the fruits. Two fruits have developed from flowers in the same whorl, and the axis extends beyond as a thin string. Compare the distance between flower scars on the stalk with the internode lengths in infertile axes in (*a*), and notice that the scars are single not triple.

facing p. 251

PART TWO

Towards a mechanism

15. Shortcomings of the proposed mechanisms

Tweedledum looked round him with a satisfied smile. 'I don't suppose', he said, 'there'll be a tree left standing, for ever so far round, by the time we've finished.'

Carroll

I have tried in the first part to expound what is certainly known, to weigh and assess what is doubtful, and to reject what seems wrong among the experimental facts of translocation. It is plain that some facts are sure and consistently repeatable: the magnitude of the dry-weight transfer, the generality of sucrose, the unique role of sieve elements and their special sensitivity to interference. Other facts are most unsure: all observations and attempts to find what modifications of protoplasts the sieve elements contain; whether anything of a visible size and nature (solution or cytoplasmic structure) may be said to be moving within or between sieve elements or with what speed. The first set of facts needs little more precise experimentation, repeating the experiments will only tell us the same things over again; the second needs concentrated and careful attention by new, imaginative methods. Between the two extremes there is a large group of facts mostly about the behaviour of radioactively-labelled molecules which require thought, arrangement, ordering and relating as well as further experimental sifting. (Some I believe are irrelevant and misleading.) Many details are known of patterns of tracer spread, of the effects of temperature, of phloem respiration, but it is not certain or widely agreed which of those data are fully reliable, which details are relevant to the main problem, which sufficiently precise and which need more dissection. At every stage of accretion of this information people have tried to use it to build models that would fit the facts. Either having insufficient facts they have devised a wrong model that later facts discounted, or having unreasoning faith in a particular model, have selected the facts that fitted it and ignored, rejected or attempted to discredit those that did not. It may be that the present stage of the discovery of the facts is no more appropriate to fruitful theorising than the previous stages; that the facts available now are not sufficient or precise enough to construct the final model. Nevertheless two things can be done: the proposed models can be examined in the light of the facts and their values and deficiencies discussed, and the general properties of an embracing model can be induced.

Certainty about the correctness of a model comes only if it is seen not only to fit all the facts, but to make verifiable predictions about new untried properties. But while some of the data on which the inductions are based are still disputed and uncertain, there is always the danger that a hypothesis may be distrusted because it is inconsistent with some observation that turns out later to be wrong. More weight should be placed in this situation on the predictive power and range of a model than on its complete agreement with all recorded data.

1. ROTATIONAL PROTOPLASMIC STREAMING (CYCLOSIS)

The proposal that mixing within sieve elements may happen rapidly by the circulation of streaming protoplasm, and transfer from one sieve element to the next be effected by diffusion, was championed by Curtis (e.g. 1929), emphasising the dependence of translocation on vital processes and the analogies with diffusion. It cannot stand quantitative examination, for even with instantaneous mixing, diffusion across sieve plates is too slow. A mass transfer of $6 \text{ g hr}^{-1} \text{ cm}^{-2}$ across a sieve plate $5 \ \mu\text{m}$ thick requires a difference of concentration of

$$C = \text{g sec}^{-1} / D \times 5 \times 10^{-4}$$
$$= \frac{6}{3600} \times \frac{5 \times 10^{-4}}{3 \times 10^{-6}}$$
$$= 0.28 \text{ g cm}^{-3}$$

between consecutive sieve elements if the whole sieve plate is as permeable to sugar molecules as an equal layer of water.

To extend the streaming mixing through the pores as illustrated by the calculation of Mason and Phillis (page 150) does not help greatly. The rates of flow required are very large, and with the Mason and Phillis assumptions, the energy expenditure impossibly high. Besides, streaming motions of this macroscopic kind are very probably not found (Chapter 10).

On the credit side, such a model would behave like a diffusion analogue both in the steady-state transfer of sugar and in the dispersal of tracer, and exhibit the known linkages with metabolism. This model has no adherents today and will not be considered further.

2. SURFACE SPREADING

The properties of interfaces between fluid phases have several times attracted the attention of physiologists seeking an explanation for translocation. The asymmetrical molecular forces in such regions, which lead to phenomena such as the familiar surface tensions of liquids, are very powerful. Van den Honert (1932) outlined an explicit analogy in which an interface between ether and water would rapidly transport molecules placed at one end of it. An interface was prepared between the two solutions. The water layer contained buffer and an indicator that changed colour in the presence of alkali. A mixture of potassium hydroxide and potassium oleate introduced at one end of the surface spread across it and the progress of the spreading could be followed by the changing colour of the indicator at the interface. The speed of spreading at the interface is very rapid, but the total transfer of substance is very small since the spreading substance is in a layer only one to a few molecules thick. Once the surface is saturated, no further motion happens unless the molecules are removed from the other end. Mangham (1917) envisaged an even less satisfactory form of this model where the sugar spread over the surface of small spherical particles and diffused between them. Hypotheses of this class remain untestable while there is no known surface with dimensions and shape in the sieve elements to do the calculations for. Physical chemists are sceptical about the possibility of continuing transport across such surfaces produced by unloading at one end.

In its simple form the hypothesis is inadequate to explain the great range in chemical properties (and therefore surface affinities) of translocated substances and speed and profile data. The linkage with metabolism may be readily explained by the need to maintain the necessary surfaces.

As knowledge of the structures remaining in functioning sieve elements grows it has become clear that there are no large surfaces that might be interfaces of the kind needed by Van den Honert's model. Whatever is there is very finely dispersed into protein filaments of 100–300 Å diameter. Though these seem not to constitute a set of surfaces in the simple sense, at this macro-molecular level they do provide a very large surface indeed considered as a phase separate from whatever they are lying in. The surface-spreading hypothesis requires the surface energies associated with the kind of water-lipid interface that is associated with a unit membrane, rather than those that might exist around a protein filament, and such unit membranes are conspicuously absent from mature sieve elements. Nevertheless protein filaments in muscle, in flagella and plant cytoplasm are capable of motions we do not understand, and if it should turn out that P-protein filaments are able to shuffle sugar molecules along their surface, the Van den Honert theory may revive in a modern form.

3. MASS FLOW, TURGOR DRIVEN

The model originally proposed by Münch (1930) was based on flow of sugar solution through sieve tubes driven by turgor pressure from sources rich in sugar to sinks where the sugar concentration was low. This model requires:

(*a*) a concentration of osmotically active sugar (say sucrose) loaded into a semipermeable compartment in the phloem;

(*b*) a pipeline of low-resistance, sugar-impermeable cells (the sieve tubes) joining the source to the sink;

(*c*) unloading, and consumption or conversion to osmotically inactive sugar through the semipermeable membrane at the sink;

(*d*) a sink for water at the downstream end. Münch saw this as a return pathway for the water from the sink end via the xylem to the leaves where it would be lost in transpiration or recirculated with more sugar.

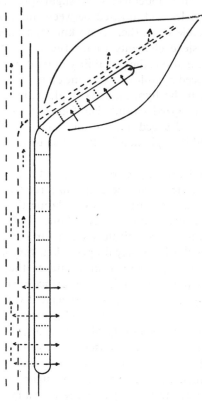

Fig. 15.1. Diagram of the source leaf, phloem and xylem as envisaged by the Münch pressure flow hypothesis. The continuous arrows show the movement of sugar as it is loaded at the source into the sieve tubes and unloaded from them at the sink. The dashed arrows show the movement of water where it does not accompany the sugar, in its return trip via the xylem to the source leaf. Sieve plates are shown dotted.

(*e*) sufficient gradient of pressure to drive the necessary flow through the resistance of the sieve tubes.

Fig. 15.1 represents the working of the model. In its simple form the model is untenable because the necessary gradients of turgor pressure do not exist. The turgor pressure gradient in the plant body is dominated by the xylem tension and the associated cells of the water pathway from high water potentials at the base to low water potentials in the leaves. It is not conceivable that the phloem cells, lying so close to the xylem, should have a reversed gradient of water potential with greatest potential in the leaves and of the magnitudes required by Fig. 12.1 to drive simple flow through 0.8-μm radius pores. Nevertheless, modified forms of the model have many staunch champions and its details must be examined. Requirements (*a*) for a sucrose-loading pump into the phloem, and (*c*) for unloading, are necessities for almost any model and fully consistent with what little is known about the ends of the system, as outlined in Chapter 3. These are fruitful fields for experimentation but not crucial to model-picking.

Requirement (*d*), the equilibration of water, will readily occur along the large gradients of water potential in the plant. The special demonstration by Münch of drops of water exuding from stripped bark (1930) and the doubts cast on these (Weevers and Westenberg, 1931; Schumacher, 1950) do not seem relevant.

Requirement (*b*), the pipeline: the sugar-impermeability is readily explained, despite the loss of the tonoplast from sieve elements, by the parietal protoplasm and plasmalemma that continue to clothe the walls when the rest of the cell contents have been converted to the specialised filamentous structures of the lumen. It is in the maintenance of this layer that adherents of this model see the connection of translocation with metabolism. For turgor-driven mass flow there is no other reason why respiratory poisons or chilling applied locally should have any effect. But they might interfere with this last vestige of living structure containing the flow matrix of the lumen. For other metabolically-driven mass flows there is a second point of influence (see below).

It is the low-resistance requirement of (*b*) which is so unsettling to any form of mass flow. Fig. 12.1 shows that the sizes of the pores in sieve plates seem to have been evolved independently of any requirement for low resistance to flow. Though the largest pores are of a size that offers very slight resistance to flow, the great majority of angiosperm sieve pores require considerable expenditure of energy for flow through them, and gymnosperm sieve areas offer very high resistance indeed. This is so if the pores are unobstructed. If the pores are further filled with filaments as in Figs. 10.5, 10.11, 10.12, the resistances are even greater, as calculated by Weatherley and Johnson (1968). Indeed there seems to be no reason for a flow system to retain crosswalls at all, even with holes in them, unless the plates take some positive part in the pumping.

There are two main reasons for holding to a flow hypothesis in the face of the resistance facts. The first is the kind of agreement noted in Chapter I between measured specific mass transfer rates and the values on the right hand side of equation 1.1. With a generalised speed figure of, say, 70 cm hr^{-1} and sieve-tube contents of 10 per cent sucrose solution, the multiplication gives

$$70 \times 0.1 = 7g\,hr^{-1}\,cm^{-2} \text{ sieve tube}$$

for the mass transfer, in rough agreement with the value of Chapter 11. This has been emphasised again by Zimmermann (1969) though he still has not made independent measurements of all three quantities, but uses Münch's figures quoted in Table 1.1. Agreement, though suggestive, does not constitute proof of the model. The second reason is the visible presence of flowing solution welling from cut bark and severed aphid stylets, though the former is very transitory, and both may be abnormal. The possible origins of aphid-stylet sugar and water have been considered in Chapter 8. There is no doubt whatever that flow of sieve-element contents will be easy for any system with sieve pores larger than about 2 μm radius, and will be inevitable in these when there is a gradient of pressure along the phloem whether caused by cutting, aphid puncture or local changes in turgor. Flow may be an important normal component of sugar movement in plants that have such pores. But equally there can be no doubt that it cannot be of universal occurrence and must be absent from gymnosperms.

(1) The prime deductive inference from this or any straight-forward flow model is that the moving contents of a sieve tube are all going in one direction, and therefore that translocation ought to be mainly one-directional whether of labelled or unlabelled material. The large amount of data which suggests that this is not so has always required special excuses. For many years refuge was sought in the suggestion that different vascular strands were specialised for transport in different directions. Driven from this covert by the improved resolution of such demonstrations as Biddulph and Cory (1960) and Eschrich (1966), for a time the adherents of mass flow could still argue that different sieve tubes were specialised for transport in opposite directions. But even this hope is now vain since Trip and Gorham (1968a) showed that ^{14}C-assimilate and ^{3}H-glucose move in opposite directions in the same sieve tubes. A position that was never logically very sound has become quite untenable.

(2) A deduction that some have made from the flow model is that everything in the sieve tube should move at the same speed, but this is

pressing the analogy too far. It is true that for flow in a glass tube the mean speed of the solvent is the same as the mean speeds of regions of solute concentrations, and that the solutes will disperse differently only in the proportions of their molecular diffusion constants (equation 13.3), but the walls of the translocation pipe are not equally impermeable to all solutes and to water. With different leakage rates for different solutes, the ratios of concentrations of translocated solutes will change with distance, the advancing profiles will change differently in shape, and the speeds, however assessed, will appear different. So a difference of speeds for different solutes does not contradict a flow model. Anyway the precision of speed measurement is low, and such experiments are unlikely to yield a certain decision on equal or unequal speeds. The most precise comparison of concentration ratios at different distances is meaningless in the presence of differential leakage. Several have gone to the lengths of trying to test whether the water moves at the same speed as various solutes (Biddulph and Cory, 1957; Gage and Aronoff, 1960; Choi and Aronoff, 1966; Trip and Gorham, 1968*b*). Now this is really an extraordinary pastime for a physiologist. It has been argued that sieve tubes may be fairly permeable to sucrose, but they can be plasmolysed; their permeability to water, if they are like other plant cells, must be around 10^5 times greater. The thought that labelled water should travel in constant proportion with labelled sugar through a pipe formed of cell membranes, protoplasm and extracellular matrix seems too fantastic to be entertained. With water all round the pipeline to exchange with, with the low water potential of the xylem within a fraction of a millimetre, the only barrier to rapid change relative to the sugar might be the fact that the water is tritiated and therefore very different in properties from ordinary water, a point neglected by the experimenters, and which would invalidate the conclusion. Tritiated water would then be behaving as a solute not a solvent. Further, as was shown on pp. 103-5, there is now experimental evidence that the water moves independently of the solutes. The only workers to consider the exchange of labelled water in such experiments are Choi and Aronoff (1966). They make the comparison between diffusive advance of tritiated water down the whole petiole, and the model of mass flow with reversible loss (model 2, Chapter 13) and find their data much closer to the former unless only a very small fraction of the phloem is carrying the translocate.

(3) The kind of profile of advancing tracer which would be predicted for a flow model can be deduced from the general considerations of Chapter 13. With rapid irreversible loss, the semilog straight line of model 1 will be formed, and with reversible loss the curves of model 2 just referred to. If the leakage of tracer solute is low, however, the dispersion

of solute as it moves will be a consequence of the Taylor equation (model 6) or of different tube sizes (model 7). From Equation 13.3, the development of an error curve in model 6 will be described by an apparent diffusion coefficient k,

$$k = \frac{a^2 U^2}{48D}.$$

Inserting the values $a = 10^{-3}$ cm for sieve-tube radius, $U = 2.78 \times 10^{-2}$ cm sec^{-1} for velocity, $D = 3 \times 10^{-6}$ cm^2 sec^{-1} for the molecular diffusion coefficient of sugar,

$$K \doteqdot \frac{10^{-6} \times 7.6 \times 10^{-4}}{3 \times 10^{-6} \times 48} \doteqdot \frac{7.6 \times 10^{-4}}{1.44 \times 10^2} = 5 \times 10^{-6}.$$

This is about three to four orders of magnitude less than the measured translocation coefficients. The presence of sieve plates in the tubes will, as has been shown, tend to reduce this value still further. The flow model cannot generate the measured profiles by this means.

In fact for simple flow the profile of advancing tracer must be dominated by model 7, the dispersion produced by tubes of different diameters. In such a system the dispersion increases rapidly as the flow advances (Fig. 13.7); half-width increases linearly with distance. Whether the dispersion is governed by tube diameter, or by the diameter of pores in sieve plates, mass flow through a bundle of sieve tubes must be determined by the variations in size of the tubes. Dispersion due to this cause must far outweigh the very small dispersion of model 6. Neither the shape of the curves nor their very rapid increase of half-width is consistent with the experimental results.

4. MASS FLOW, ELECTRO-OSMOTIC

The clearest objection to mass flow is the presence of the sieve plates. Spanner (1958) proposed to retain some of the attractive features of mass flow while using the sieve plates as a site of pumping. In this way, instead of a pressure gradient falling progressively at each sieve plate, some energy input would make a steep positive jump at each sieve plate, perhaps more than compensating for the fall in pressure along the sieve element (Fig. 15.2).

The mechanism he proposed for this energy input was electro-osmosis. This is the phenomenon that occurs when water permeates a charged membrane, and an electric potential is maintained across the membrane. With a negative charge on the membrane, the solution flows towards the

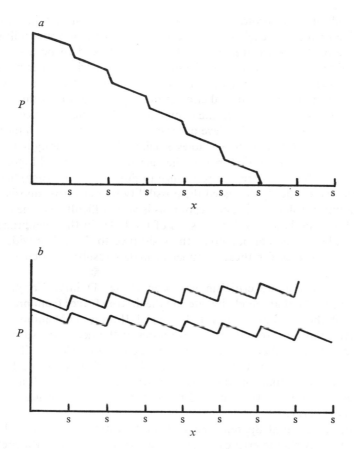

Fig. 15.2. Changes in pressure (P) with distance (x) along a sieve tube on flow hypotheses. Sieve plates occur at the points s. In both a and b the resistance to flow in a sieve element causes the same pressure drop between sieve plates. In a, the sieve plates offer a further and greater resistance to the flow, and the pressure drop across each plate is substantial. In b, by some active mechanism, the pressure gradient is restored partially (lower curve) or to a higher value (upper curve) at each sieve plate, and the overall pressure gradient either declines much more gradually or increases with distance. After Spanner (1958).

negative potential. The phenomenon may be considered as an extension of electrophoresis: a suspension of negatively charged particles, say macerated filter paper, migrates towards the anode. If the particles are held still, as in a piece of filter paper, the same forces cause the water to flow towards the cathode. Spanner pictured each sieve plate with a potential maintained across it polarised in the same direction. The sieve-plate wall and any protoplasmic structures will almost certainly carry a negative charge, so each sieve plate would need to be some millivolts-negative on the downstream side. There would need to be a means of preventing a short circuit along the sieve element, and another means of maintaining the potential across each plate. The latter is achieved in Spanner's model by a pumped circulation of potassium ions through the sieve plates and a return via the companion cells. The presence of cytoplasmic components in the sieve pores was seen as enhancing rather than restricting the flow: with the holes smaller, Spanner thought that the enhanced electro-osmotic forces in the narrower spaces would more than compensate for the increased resistance to flow. Fensom (1957) had earlier and less explicitly discussed the possibility that electro-osmosis might be driving translocation. The hypothesis is very difficult to come to grips with. So much is uncertain: the sizes of the holes in the membrane; the magnitude of the zeta-potential; the resistance to flow. The wide range of guessed values for these parameters makes resulting calculations of little value.

One more definite criticism has been made (Dainty, Croghan and Fensom, 1963), that merely by altering the driving forces from pressure to electrical, the necessary work is not changed. If there is not enough energy to drive solution through small sieve pores at the necessary speeds (Fig. 12.1) by pressure, then neither is there enough to do it electrically. Indeed Dainty *et al.* show that the electro-osmotically driven flow must always use more power than pressure-driven flow because there will always be greater slip between the ions and the water and therefore greater dissipation of energy.

A deduction that Spanner makes from this hypothesis, and which is more strongly characteristic of this form of mass flow than of the pressure-driven kind, is that translocation will be almost independent of distance. With the driving force located at each sieve plate, the flow is sustained all along the pathway. 'The total available effort, unlike the state of affairs in the pressure flow hypothesis or for that matter the cohesion theory of xylem transport, increases proportionately to the path length and the problem of transport is no more serious for a tall plant than a short one.' This little dependence on distance sharply distinguishes flow hypotheses in general, and this version in particular, from the accelerated diffusion one to be considered next.

5. ACCELERATED DIFFUSION: MOVING STRANDS
The intracellular strands described by Thaine in sieve tubes stimulated me into advancing a hypothesis which has many interesting properties.

Though the reality of these macroscopic strands is now doubtful, and no trace of them has been found in electron images, the hypothesis does not depend solely on their presence and can be adapted equally to make use of the plasma filaments that are a constant feature of recent electron images of sieve elements. The model was described qualitatively (Canny, 1962*b*) and later its quantitative behaviour was analysed (Canny and Phillips, 1963). The model comes so close to explaining the totality of experimental facts that the failure of its structural basis prompts me rather to seek a similar structural framework of different size than discard the hypothesis. The model will be described in its original form and its properties explained. Later, a revision of the structural framework will be suggested.

The sieve elements are pictured as containing two phases: a stationary sap phase that is essentially just a solution of sugar in water; and a moving cytoplasmic phase that streams in both directions through the sap in long strands traversing the sieve element, penetrating the pores of the sieve plates and extending into the next consecutive element of the sieve tube. Some strands stream in one direction, others in the reverse direction. The whole sieve tube is seen to be something like an elongated cell whose trans*vacuolar* strands have become enormously extended, and are supported by the sieve plates, rather as insulators on a telegraph pole keep the wires apart. There is exchange of solute (sugar for the present, but the model will transport almost anything) between the strands and the sap at a rather slow rate, governed by the permeability of the strand surface to the sugar. The motion of the cytoplasmic strands is maintained like protoplasmic streaming by the respiratory activity of the sieve elements themselves and perhaps also the companion cells and surrounding parenchyma cells, using as substrate the sugar of the sieve-element sap.

Such a model operates just like a diffusion system, but with an increased value of the diffusion coefficient. If the sugar concentration is the same all along the sieve tube there is no transport of sugar. The strands move, the sugar exchanges between them and the sap, but no net movement occurs. If the sugar concentration is higher at one end of the tube, the strands moving down the concentration gradient carry this higher concentration with them, slowly releasing it to the sap, while the strands moving up the gradient carry a smaller sugar concentration than the sap through which they pass, and slowly gain sugar from the sap. The effect at an intermediate sieve element is to (greatly) steepen the gradient of sugar concentration between the down strands and the up strands. The down strands have a high sugar concentration from further back up the concentration gradient; the up strands have a low sugar concentration from further down the gradient; the sap has an intermediate concentration (Fig. 15.3).

The steeper the concentration gradient along the sieve tube the greater the net transport of sugar. The strands must have enough structural rigidity to resist deformation by the osmotic forces set up by the different concentrations of water in the various zones, or else the differences in concentration would be ironed out by movements of water.

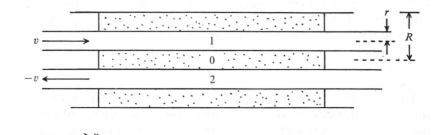

Fig. 15.3. Simplified model of the streaming strands. Only two strands are represented: a strand moving to the right (positive) with velocity v, where sugar concentrations are given the suffix 1; and a strand moving to the left, velocity $-v$, where sugar concentrations are given the suffix 2. In the sap, concentrations are given the suffix 0. Distance is positive to the right. The radius of each strand is r, and that of the sieve tube, R.

Putting this in algebraic terms: with n strands in the sieve tube each of radius r, half the strands moving one way and half the other; the concentration of sucrose (or other solute) is represented by s with a suffix to show the region concerned. The concentration of sucrose in the strands moving to the right, the positive direction, is s_1, in those moving to the left is s_2 and in the stationary sap is s_0.

Two other parameters are necessary: the permeability of the strand surface to solute, $\gamma (\text{cm sec}^{-1})$ and the ratio of strand area to the total area of sieve tubes ρ. The latter is conveniently set at 1 (strands occupying half the sieve tube) since Esau and Cheadle's (1959) measurements showed the sieve plate to be about half occupied by pores. The equations for the concentrations s_1, s_2, and s_0 contain only these variables:

$$\frac{\partial s_1}{\partial t} + v \frac{\partial s_1}{\partial x} = \frac{2\gamma}{r} (s_0 - s_1) \tag{15.1}$$

$$\frac{\partial s_2}{\partial t} - v \frac{\partial s_2}{\partial x} = \frac{2\gamma}{r} (s_0 - s_2) \tag{15.2}$$

$$\frac{\partial s_0}{\partial t} = \frac{\rho\gamma}{r} (s_1 + s_2 - 2s_0). \tag{15.3}$$

When the system is in a steady state these equations lead to simple descriptions of the transport process.
First:

$$s_0 = \tfrac{1}{2}(s_1 + s_2).$$

That is to say that everywhere along the sieve tube the sap concentration is equal to the average concentration in the strands.
Second:

$$s_1 = s_2 + \text{constant}$$
$$= s_2 + \sigma, \text{ say}.$$

So the difference in sucrose concentration σ between the positive and negative strands is the same all along the sieve tube. If S_1 is the sucrose concentration at the top of the gradient where $x=0$ in the strands moving in the positive direction, the concentrations in the three regions all decrease linearly with distance as shown in Fig. 15.4. The slope of these lines is $-\sigma\gamma/vr$ per unit length.

The net transport of sucrose along each sieve tube is

$$T = \tfrac{1}{2}n\pi r^2 (s_1 - s_2) = \tfrac{1}{2}n\pi r^2 v\sigma \qquad (15.4)$$

since $\tfrac{1}{2}n$ strands, each of area πr^2, are moving in each direction. T is equal to a constant (the translocation coefficient, K) multiplied by the concentration gradient, and K can be readily seen to be equal to

$$\frac{\rho r v^2}{2(\rho + 1)\gamma}.$$

Or, putting $\rho=1$, $K=rv^2/4\gamma$. This expresses the translocation coefficient simply in terms of three properties of the model, and shows that more efficient transport is obtained by (a) having larger strands, (b) having faster strands (most effective since $K \propto v^2$) or (c) having less permeable strands. It can be seen in a general way that such a model will account for a great many of the experimental facts: the close analogy with diffusion both for ordinary sugar and for tracer molecules; the transport of different substances, since anything that gets into the strands and to which they are not too permeable will be transported; simultaneous bi-directional transport in a single sieve tube, since net movements are in the directions of the gradients. Further, the model can be seen as an extension of known cellular structures, and its evolutionary development required only the co-operation of neighbouring transvacuolar strands through the sieve pores. But beyond all this the model makes quite specific quantitative predictions which will be the subject of the next chapter. Many of them fit remarkably well with the experimental data.

The prime objection to the model is that the strand structures it uses

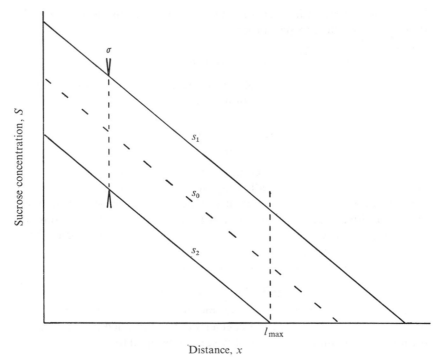

Fig. 15.4. Results of the steady-state analysis: the variation of sugar concentration with distance in the three regions. Sugar concentrations all have the same gradient, and the difference in concentration between positive and negative strands is a constant σ, at all points. The concentration in the reservoir is half-way between that in the two sets of strands. At the distance l_{max} the concentration in the negative strands is zero, but those in the reservoir and the positive strands are $\sigma/2$ and σ respectively.

seem not in fact to be there, nor any such tonoplast-type membranes as were originally imagined as separating them from the sap. Instead the filamentous contents of the sieve elements appear to be much smaller. There are however, apparently still two phases and the P-protein filaments could well be made of a contractile protein and could be moving. Further, it will be seen below that the experimental data point to the conclusion that the strands must be smaller than Thaine-strands, and of a size in the range where $r \doteqdot 100$ Å. There are some other inconsistencies of a detailed and complicated kind that will be brought out in the next chapter.

6. CYTOPLASMIC PUMPING

In its first-envisaged form the strand-diffusion model relied on protoplasmic streaming to move the oppositely-directed streams. More recent studies of streaming protoplasm in *Nitella* and other cells show that in the regions of protoplasm where the forces originate there are long plasma filaments about 70 Å in diameter, thought to consist of contractile protein, and likened to the filaments of actin in muscle. The filaments of phloem protein are substantially thicker and have a more obvious fine structure but they could also be made of contractile protein and could generate motions of fluids in the sieve tubes. Such motions could be of many kinds, and one of these has been made specific by Thaine (1969). He suggests that contractile protein of the sieve tubes is organised in rings in the outer sheath of the old transcellular strands. These contain endoplasmic tubules packed inside in which the sugar solution flows. The longitudinal fibres are again continuous from one element to the next through the pores. Flow of sugar solution in these endoplasmic tubules is generated by peristaltic pumping as a wave of contraction and relaxation travels along the strand sheath.

This concept has been further investigated by Aikman and Anderson (1971) who apply a recent theory of peristaltic flow in tubes to a system of the dimensions of transcellular strands, but dealing with only a single tubule, not with the parallel array of Thaine's drawing. With 1-μm radius strands, 10 per cent sucrose solution and a contraction wave that has a velocity 5×10^{-2} cm sec^{-1}, wavelength $100\ \mu$m and relative amplitude 0.4, the transfer of sucrose achieved within a tubule is 8.6 g hr^{-1} cm^{-2} tubule.

Thaine does not make it clear whether his strands are pumping all in one direction in a sieve tube, but Aikman and Anderson consider the possibilities of oppositely-directed tubules in different bundles, different sieve tubes, or within a single sieve tube. From the above calculated flux, to achieve the required specific mass transfer rates of 5 g hr^{-1} cm^{-2} phloem it would be necessary for all tubules to be pumping one way in rapidly translocating systems, but for systems requiring less net transport, tubules working in opposite directions would provide precisely similar bi-directional and diffusion-analogue kinetics to the Canny/Phillips model. It is possible that the direction of pumping can be controlled by the overall sugar gradient, conforming to the Mason/Maskell rule.

The energy dissipation of the pumping system is calculated and the rate of working of the tubule wall shown to be about 0.5×10^{-3} J m^{-2} sec^{-1}. We may translate this into the terms of Chapter 12 by finding the surface area of tubule in a cubic centimetre of phloem. Take, for simplicity, half the phloem area occupied by tubules (as would be approximately needed from the above mass transfer values). Then the number of tubules in a cubic centimetre of phloem is $0.5/\pi \times 10^{-8} = 1.6 \times 10^{7}$, and the surface area of tubule in a cubic centimetre,

$$1.6 \times 10^7 \times 2\pi \times 10^{-4} \text{ cm}^2 = 1 \text{ m}^2.$$

The rate of work is then

$$0.5 \times 10^{-3} \times 3600/4.2 \text{ cal hr}^{-1} = 0.48 \text{ cal hr}^{-1},$$

very close to the 0.5 cal g^{-1} fw phloem hr^{-1} which is probably available (page 162).

A third check on the feasibility of the model is achieved by showing that the necessary tension in the wall sheath could easily be developed by contractile elements similar to those of vertebrate muscle. However the rate of contraction required by the assumed values is half as fast again as that of frog sartorius muscle, and 100 times that of invertebrate smooth muscle.

This model has many basic similarities to the Canny/Phillips model, and if the tubules are oppositely directed and slightly permeable to sucrose, differs only in the imagined motive power.

16. Useful properties of the accelerated diffusion model, and some difficulties

The inventor produces his invention because, in the existing circumstances, it has become possible. It is true that he, himself, tends to confuse the issues by persuading himself optimistically that his invention has real and important utility ... But this is mere self delusion. In reality, the inventor is concerned with his invention.   R. Austin Freeman

I. MASS TRANSFER, GRADIENT AND TRANSLOCATION COEFFICIENT

The first requirement of any model is that it shall adequately transport the sugar in the amounts outlined in Table 1.1. To assess whether this is possible by the activated diffusion model is no simple task. It is useless to make the kind of naive calculations given by Weatherley and Johnson (1968) about the amount of sugar the strands carry through the pores. The model is a diffusion analogue and must be treated as such. The real question is, what value of the translocation coefficient K can be expected from the model?

If a value around 5×10^{-2} cm^2 sec^{-1} can be expected, then a specific mass transfer of 6 g hr^{-1}/cm^{-2} sieve tube can be achieved with a sucrose gradient in the sieve tubes of

$$6/(5 \times 10^{-2} \times 3600) \text{ g cm}^{-4} = 3.34 \times 10^{-2} \text{ g cm}^{-4}$$

which is 10.6 M sucrose per metre or 1 M per 9.4 cm. (Refer to Table 3.1 for other gradients necessary to achieve selected transfers, remembering that these are on the phloem-area basis.) I am prepared to believe in a gradient of this magnitude as the upper limit for the fastest and short-period translocation over short distances. The smaller values of Table 1.1 are easily accounted for by sugar gradients of 1 M per metre or less.

Now values of K of around 5×10^{-2} are a measured property of the system both derived by Mason and Maskell and as measured from the error functions produced by advancing tracer. Analysis of the model (Canny and Phillips, 1963) showed that operating in the steady state with no concentration changes with time, and with half the lumen filled with strands, the translocation coefficient is given by

$$K = rv^2/4\gamma \tag{16.1}$$

where r is the strand radius, v is the strand speed, and γ is the permeability coefficient of the boundary between sap and strand. All three quantities are unsure, but various assumptions can be made:

(*a*) If we assume Thaine-type strands of radius 1 μm and moving at speeds of visible protoplasmic streaming (10 cm hr^{-1}) we have

$$5 \times 10^{-2} = \frac{10^{-4} \times (2.8)^2 \times 10^{-6}}{4\gamma}$$

and the permeability of the strand surface would have to be around 5×10^{-9} cm sec^{-1}, intermediate between free diffusion (10^{-6}) and a tonoplast-type membrane (10^{-11}).

(*b*) Assuming rather that the strands are of the size of P-protein filaments (200 Å diameter), and their speeds those of the advance of tracer profiles (2.8×10^{-2} cm sec^{-1}), we have now

$$\gamma = \frac{1.0 \times 10^{-6} \times (2.8)^2 \times 10^{-4}}{4 \times 5 \times 10^{-2}} = 5 \times 10^{-9} \text{ again.}$$

Neither assumption may be fully satisfactory, but both would work in the sense that they would sufficiently accelerate diffusion to produce the mass transfers of Table 1.1, given sucrose gradients like those in Table 3.1.

So on either one of two reasonable sets of dimensional assumptions the model leads to a sufficient acceleration of diffusion to provide values of K comparable to the measured ones. In each case the boundary between moving strand and stationary sap needs to have about the same permeability to sucrose, and this permeability is intermediate between that of free diffusion and that associated with a unit-membrane type structure. It was a paradoxical feature of the Thaine-sized strand version of the model that this membrane permeability had to be much higher than that of a unit membrane, while the only boundary that might conceivably be separating protoplasmic strands from a watery sap would be a tonoplast. The special modifications of the tonoplast needed to increase its permeability 1000-fold seemed difficult to envisage. With the smaller plasma filaments, however, no such unit-membrane boundary is seen or expected, and until something happens to prove me wrong I am content to believe that they may have a sucrose permeability of about 5×10^{-9} cm sec^{-1}.

The question of whether adequate gradients of sugar concentration exist to take advantage of the high value of K and lead to rapid mass transfers in some systems is one which cannot at present be answered. They were there in Mason and Maskell's cotton plants, first revealing the importance of K. But neglect of these ideas has meant that few workers have been interested to measure longitudinal gradients of sugar in the sieve tubes, and there are only those few measurements from trees quoted in Chapter 3. These gradients are of gentle slope (0.02 M sucrose per metre and would produce only very slow transfers of dry weight down the whole tree trunk, a point to be taken up again when the effect of distance is considered. The specific mass transfer may be estimated from such a gradient and a value of K of 5×10^{-2} cm^2 sec^{-1} as

$$
\begin{aligned}
\text{SMT} &= 0.02 \times 10^{-5} \times 5 \times 10^{-2} \text{ moles sucrose sec}^{-1} \text{ cm}^{-2} \\
&= (0.02 \times 5 \times 3.6 \times 10)/10^5 \text{ moles sucrose hr}^{-1} \text{ cm}^{-2} \\
&= (0.02 \times 5 \times 3.6 \times 3.6 \times 10^3)/10^5 \text{ g sucrose hr}^{-1} \text{ cm}^{-2} \\
&= 1.3 \times 10^{-2} \text{ g hr}^{-1} \text{ cm}^{-2}.
\end{aligned}
$$

This is 100-fold less than the transfers into fruits, etc. The problem has nothing specifically to do with the activated diffusion model, but is a product of thinking about the transfers in terms of a diffusion analogue. To assess the worth of these ideas there is urgent need of more measures of sugar gradients, particularly in such organs as fruit-stalks where the transfers are large. These measurements could profitably be coupled with simultaneous tracer measures of K, and mass measures of the specific mass transfer.

2. TRANSLOCATION, DISTANCE AND MORPHOLOGY

A deduction from the activated diffusion model that was made in the original analysis is that the maximum rate of transport is limited by the distance over which the solute is being transported. High rates of transport can be achieved only over short distances; over long distances only low rates are possible. This follows from Fig. 15.4, where l_{max} is marked as the distance at which the concentration s_2 in the negatively directed strands becomes zero, since s_2 cannot be negative. The slope of the lines is known to be $-\sigma\gamma/vr$, so

$$
l_{max} = \frac{(s_1 - \sigma)}{\sigma\gamma} vr. \tag{16.2}
$$

At this distance the concentration in the sap is $\frac{1}{2}\sigma$ and in the strands arriving, is σ. Equation 16.2 can be re-arranged to relate l_{max} to the rate of transport per unit area of sieve tubes, τ. Then, for $\rho = 1$,

$$
l_{max} = \frac{(vs_1 - 4\tau) vr}{4\gamma\tau}. \tag{16.3}
$$

This is a very serious limitation of any translocation model and needs careful consideration. It is unique to this kind of model, the diffusion analogue. It is not shared to the same extent by simple mass flow, where the rate of transport falls linearly with distance. Electro-osmotically driven mass flow is little attenuated by distance, or may not be so at all. Quantitatively it is not easy to set sensible bounds to the kinds of limitation this effect would impose on the logistics of the plant body while we are in ignorance of several of the values in Equation 16.3. It can be said in broad terms that high rates of transport could be maintained over centimetres or even decimetres, but over metres of phloem the rates would necessarily be low. An illustration can be attempted in rough quantitative terms.

It has been shown above that two quite different sets of assumptions about the size and speed of the strands lead to the same estimate of γ, the permeability. Using these figures again:

$$v = 2.8 \times 10^{-2} \text{ cm sec}^{-1}$$

$$r = 1.0 \times 10^{-6} \text{ cm}$$

$$\gamma = 5 \times 10^{-9} \text{ cm sec}^{-1}$$

and putting s_1 at 0.25 g cm^{-3}, the graph of maximum transport rate against distance shown in Fig. 16.1 is obtained. There is an error of 10 in the ordinate of the version of this curve in Canny and Phillips (1963). This may be taken as representing translocation systems in general, though different plants will differ in the values of the constants and have slightly different curves. It seems that the highest values of the specific mass transfer τ can be expected only over distances less than about 5 cm; that up to about 30 cm there is no difficulty in maintaining a specific mass transfer of 1 g hr^{-1} cm^2, but that over distances greater than 3 m, τ is down below 0.15. Beyond this distance it is decreasing only very slowly with extra path length.

This may seem a drastic limitation, but a recollection of the shapes and dimensions of plants may lead the reader to share my conviction that plant bodies are organised to operate under some such rule, that the pattern of sources and sinks shows signs of having evolved in co-operation with a transport system that could not cope with rapid transport and long distance simultaneously. Consider how the rapidly growing fruit, the pumpkin, the corn cob, the coconut, the bunch of grapes, is always subtended by a large leaf or cluster of leaves close by, and from which it is known the sink draws nearly all its substance; how the removal of this source, the close leaf(-ves) severely limits the growth of the fruit, although and because the supply is now switched to a remoter source. That successful higher-plant unit the short shoot, which has evolved independently in

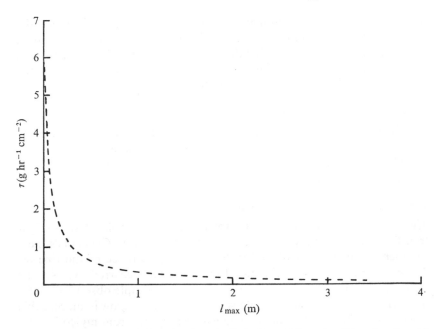

Fig. 16.1. Theoretical relation showing the limitation of the rate of translocation (τ) by path distance (l_{max}) on the accelerated-diffusion model. For the assumptions inherent in the calculation, see text.

many groups, is the prime example: the compact cluster, almost independent from the rest of the plant, of a whorl of leaves, a few flower buds fed from the leaves and producing one or two substantial fruits. A great many of the fruits we eat are produced in such systems. It is not an accident that the cereal grasses have sessile florets.

Conversely plants that produce fruits remote from leaves on much-branched inflorescences, tend to have small fruits: the umbellifers, the non-cereal grasses, the Polygonaceae, the Chenopodiaceae, the sedges. Active sinks that grow quickly underground, like yams and potatoes, do so not far from a cluster of leaves close above the surface of the ground, they do not develop at the foot of a tall leafless stem. Sinks that do so, like artichokes and dahlia tubers, are small and slow growing, unless there are leaves near the ground or unless the stem acts as a half-way house, accumulating reserves over a long period of assimilation, and then supplying them rapidly over a shorter distance during a burst of tuber growth. Incoll and Neales (1970) show that this is just what happens in the artichoke. Unfortunately they did not record the rates of depletion of reserves from different stem heights, but this hypothesis predicts that the rate of depletion would be most rapid from the lower parts of the stem. Here is a physiological basis for the empirical rule induced in Chapter 5, that sinks are supplied by the nearest sources. A deduction from the model independently produces a rule that experimental observation had suggested.

There are two apparent exceptions to this rule, and since they contradict the whole basis of the otherwise-consistent hypothesis, they need to be looked at closely. The first is the sausage-tree fruit and the second is the tall tree trunk. In both systems there seems to be rapid transport over a length of phloem.

The problem of the sausage tree

The sausage tree was used in Chapter 1 as an example of dry-weight transfer, and the specific mass transfer into the fruit in Clements's measurements is in no sort of doubt. From the appearance of the ripe infructescence it seems that this rapid transport has occurred over 50–100 cm of fruit stalk (Plate 23). This is a matter that simple observation of a number of fertilised flowers would settle, but the trees grow in inaccessible places, and I have not so far had conclusive reports from my colleagues who have access to them. I have myself examined the fruits only at the fully ripe stage when each is indeed suspended on a long thick stalk far from the crown of leaves that has produced it. But there seems to be evidence that the stalk elongates after fertilisation; that the rapid growth of the fruit occurs when the fruit is much closer to the leaves. It will be seen in Plate 23 that the thick fruit stalk bears a slender long branch distal to the insertion of the fruit. This is the old inflorescence axis, and on it can be seen the scars of the old unfertilised flowers. The thick fruit stalk is produced by the very large increase in diameter of (*a*) the inflorescence axis up to the node of the fertilised flower, (*b*) the peduncle and of the flower, (*c*) part of the floral axis. This thick, straight but still two-kneed stalk is straightened into a single rope-like suspension by the great weight of the fruit. I believe that it sometimes also elongates, for two reasons: (1) the flower node scars on the thin dry part of the inflorescence axis are much closer together than the scars on the thick part above the fruit (Fig. 16.2); (2) the following sentence in Clements's (1940) account: 'By far the greatest increase in fruit volume is accomplished during the first five weeks, for it is during this period that the largest daily increments are observed. During this same period the daily increments of dry material are higher than those observed after the floral stem is more mature. During this early period, the tissues of the floral stem are young; in fact the long stem (145 cm), which for the most part is really a single

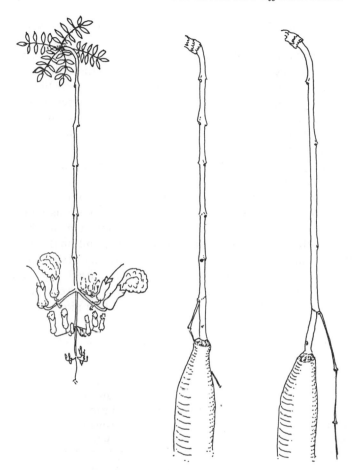

Fig. 16.2. Suggested alternative forms of the mature fruit stalk of *Kigelia*.

At the left is the hanging inflorescence at a middle period of its life. Flowers are borne in threes on side branches which also arise in threes at whorls on the inflorescence axis. The first five floral nodes of the axis have shed their floral branches without any flower being fertilised, and triple scars remain at each node. The nodes are spaced at 15–20 cm. In the first whorl of nine existing flowers two appear to have been pollinated (bats) and have shed the corolla. Two will open soon (at night) and two have been shed. The remaining three are sketched behind. The more apical flowers are still young buds.

In the centre drawing the fruit has developed at the sixth floral node on the axis, all other flowers have been shed, and the fruit stalk has thickened without elongation, while all the scars of floral branches remain triple. The nodes are still 15–20 cm apart. The axis continues beyond the fruit.

In the right-hand drawing is shown the suggested alternative mode of development of a long stalk. A flower has been fertilised at the second floral node of the axis. The first triple node has come apart by growth at the node, and the three scars are now at different levels separated by 30–50 cm: a single node has become two internodes. The axis continues beyond the fruit in its unextended form.

internode, is still elongating.' The 145 cm refers, I take it, to the length after elongation.

The process by which such elongation may occur is suggested by careful examination of the fruit stalk. Here the scars of the fallen flowers are often single at each node (as in Plate 23), whereas the flower branches were born in whorls of three, and beyond the fruit the withered axis has three scars at each node. The elongation process seems to occur at the triple node, extending the node so that the three scars which started at the same point become separated by many centimetres. This does not occur on all axes, and may be limited to some nodes on a single axis. The different stages are drawn in Fig. 16.2.

The problem has been slightly clarified by some data kindly collected in Florida by Dr P. B. Tomlinson and Dr Stephen Manis and which I publish here with their permission. They recorded fruit growth by measuring fruit length at various times from pollination, and final fruit volume, fresh weight and dry weight at successively later harvests. They kept a record of the peduncle lengths. Two varieties were used, *Kigelia africana* (*K. pinnata*), flowering in February–April, and the other one in June–August. The latter had the greater growth rate and that alone is given here.

TABLE 16.1   *Dry weight transfers into* Kigelia *fruits* (Data of Tomlinson and Manis)

Translocation distance cm	Time from pollination to harvest days	Dry weight of fruit g	Extension of peduncle during growth cm	Calculated specific mass transfer $g\ hr^{-1}\ cm^{-2}$ ph
143	16	14.5	0	0.07
184	23	120	0	0.39
113	29	172	−2	0.47
88	36	284.5	+1	0.62
143	50	301	0	0.47
161	57	172	0	0.24
161	64	253	0	0.31

The final dry weights of the fruits are listed in Table 16.1, with the time of growth and length of translocation path from the nearest leaves. No record was kept of the phloem cross section in the peduncle, but it seems fair to use that of Clements (Chapter 1) who was apparently working on the same species in Hawaii, so a value of 0.53 cm^2 of phloem has been used to calculate values of the specific mass transfer. It is at once clear that the stalk lengths are long by the standards of Fig. 16.1; and that they have not elongated during fruit growth. This fact correlates well with the assumed mode of elongation where this occurs, for drawings of the fruit stalks show that the nodes are all triple. It is also very clear (refer to Table 1.1) that the values of specific mass transfer are much smaller than in Clements's experiments, and low by comparison with most fruits. Reference to Clements's paper will show that his dry weight increases in a week often exceeded these for the whole growth period of a month or two.

A plot of these calculated values of specific mass transfer against distance of translocation is shown as Fig. 16.3. Of course in the absence of real measures of the phloem of the stalks, and since the data derive from different times on a growth curve which is probably sigmoid, the points may well be somewhat displaced, but a general trend is apparent and it is

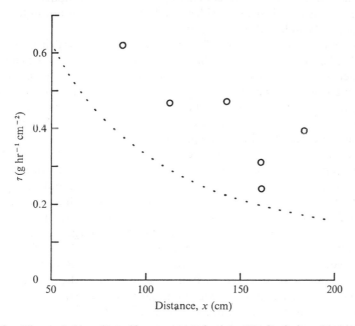

Fig. 16.3. The variation of specific mass transfer into *Kigelia* fruits with length of translocation path. The calculated values of specific mass transfer of Table 16.1 are plotted against the measured translocation distances. A constant value of 0.53 cm^2 phloem per fruit stalk has been assumed, and no adjustment for changing growth rate with time has been made except for the omission of the 16-day value. Fruit growth was clearly very slow in the early stages. The dotted curve shows the theoretical fall of transfer with distance with the assumptions made in Fig. 16.1.

even possible to persuade oneself that the trend could be the appropriate part of the curve of Fig. 16.1.

It seems reasonable to guess that Clements was working with flowers near the base of the axis, borne on shorter stalks which did elongate, and that Tomlinson and Manis would have measured higher rates of fruit growth if they had chosen shorter stalks. The question remains tantalisingly open.

The problem of the tree trunk

Kigelia is an isolated and doubtful example, but trees are everywhere. They have an organisation that clearly works, and is based on translocation however that functions. Any hypothesis that cannot explain a tree is useless. The problem has been touched on several times: that a long tree trunk can have only a very gradual gradient of sugar concentration, and an accelerated diffusion model would therefore limit it to small rates of transport overall. Before concluding that any such model is therefore nonsense it is only just to consider whether a tree might not in fact function perfectly well with this limitation.

The tree trunk is a mass of stored carbohydrate, much of it easily mobilised and transportable radially along the wood rays to the enclosing cylinder of phloem where it can move up and down. The metabolism of its living cells is slowly consuming this store, and more carbohydrate is being fed in daily at the top in the growing season. One may think of it in diffusion terms as a very large diffusion space full of solute with a high diffusion coefficient. Within this space transport may be locally very rapid. That is, if an active sink for solute arises at a point, solute will be transported rapidly to that point from neighbouring stores down steep gradients. There is thus ample transport capacity to grow a branch shoot or cauline fruit or a new root quickly at need. But the traffic would have to be local. There could be no question of a deep root tip growing with carbohydrate supplied daily from a lofty leaf. The large storage capacity of the wood is seen as the solution to the problem. Much of the stored carbohydrate may have been held there since the leaves were near when the tree was smaller; much of it may have arrived there slowly down gradual gradients at times of vigorous leaf assimilation.

Zimmermann (1969) is talking about this problem when he says, in abrupt unexplained logic that must be puzzling to a reader not versed in this controversy: 'For the first time we have directly demonstrated phloem transport over distances of 12 meters, at a peak mass transfer. This means that a mechanism as described by Canny and Phillips (1963) is entirely inadequate an explanation of translocation.' Zimmermann has made no such demonstration, he has estimated a speed of a ratio wave of concentration. He uses Münch's (Table 1.1) values for mass transfer into tree trunks.

But Münch's data are, like those of Crafts (1931a), measures of a transport across the top of the crown from leaves into the main body of trunk plus roots. They are not measures of transport down a long distance from leaves to roots. All that is required to accommodate Münch's measures to this distance limitation is that the main growth increment of a tree be at the top, within a short distance of the leaves. This is in fact so. The shape of the tree, with its thick base and tapering limbs, makes us think superficially that the main growth increment must be at the bottom where it is thickest. But this is only where it has been growing *longest*. Each season the annual rings are wider at the top, and they are spread all over the branches, making up substantially more mass (volume) increment high up than low down. 'Trees in general grow slowly in relation to their mass. Spring vegetation may be largely produced from stored reserves; summer growth is nourished from current supplies; and products of assimilation may be stored largely during the latter part of the season...Although materials must move great distances in trees, related processes go on slowly; rapid translocation probably does not occur.' (Crafts, 1938.)

There are no experimental observations on these matters because there has been no stimulus to make them. I am trying to put the view that our understanding of the functioning of a tree has been much too naive; that the general consistency of the translocation data forces me to consider that a tree may function in a more complex and subtle way; and that the time has come for someone to do some experiments.

3. TRANSPORT OF DIVERSE SUBSTANCES

An attractive feature of the model is that it will carry any substance from places where it is abundant to where it is relatively impoverished, provided only: (a) that the substance can be got into the sieve tubes; (b) that it does not disrupt the movement of the strands; and (c) that the permeability to it of the strand-reservoir boundary is less than its free diffusion in water.

As was shown in Chapter 4 a great range of naturally-occurring and exotic substances seems to be translocated by the phloem, from virus particles to fluorescein, from inorganic actions to organophosphorous insecticides. Any of these or any metabolic intermediate, present locally and penetrating the sieve tubes either by the plasmodesmata of the symplast or from the free space by way of the walls and the plasmalemma, would build up a concentration in the sap of the local sieve tubes, and would be distributed by the strands down the gradient of concentration to parts of the plant where there was less of it. Then the net transfer would cease unless the downstream concentration was kept low by unloading and use. The utility of such a transport system, carrying everything to where it is needed at the moment from the nearest source, and at a rate which increases with the demand, is readily appreciated. Indeed as a specification it reads like a bio-engineer's dream.

Any substance carried by the system should behave very much in the same way as sucrose, but differ in the permeability constant between strand and sap. The other properties of the model are not substance-specific, but for each substance one would expect a characteristic value of this coefficient (γ), and the larger its value the less would be the efficiency of transport (smaller K). Ways of measuring γ will be discussed in section 9 of this Chapter. It will be plain to the reader that a start has still to be made on collecting the relevant data for important non-sugar compounds such as amino acids, organic acids and nucleotides.

Some substances there will be which will disrupt the motions of the strands themselves, either by interfering with the supply of energy or by breaking up some necessary cytochemical organisation. The striking position of calcium as the non-translocatable substance may fit in here, and its action may prove to be correlated with its known effects on proteins and phospholipid membranes, and with its double valency. Cyanide ions may be expected to operate by interrupting the energy supply (Chapter 12) and so stop their own transport, but the system apparently recovers when they are removed.

4. PATTERNS OF MOVEMENT

The broad conclusion of Chapter 5 was the generalisation that a sink was supplied by the nearest source. What has been said in section 2 of this Chapter about the distance dependence of the accelerated-diffusion model shows how such a rule would be a necessary consequence of such a mechanism. The switching of supply to the next nearest source when the nearest is removed would also follow, as would the observed consequence that the rate of translocation from the further source is slower. The general diffusion-analogue properties of tracer movement were outlined there and left at what seemed to be the stage of a far-fetched conclusion (page 65): that tracer sucrose ought to move into the phloem of mature leaves, but would probably remain there. The hypothetical pumps moving sugar from mesophyll to sieve tubes are unlikely to allow it to spread into the inter-vein tissue. The faint tracer images often found on autoradiographs of mature leaves (page 63) might be interpreted as evidence of the presence of this small amount of phloem-limited label.

These ideas led to the prediction that aphids feeding on the mature leaves of a plant in which labelled assimilate is moving ought to accumulate label while the leaves on which they are feeding continue to show little or no label. This prediction was verified for *Aphis craccivora* feeding on the leaves of *Vicia faba* during six hours while ^{14}C-labelled assimilate was moving from an upper leaf to the apex and down the stem to the roots (Canny and Askham, 1967). Reference to the autoradiographs of that paper will show how a number of aphids are strongly labelled while the leaves on which they are feeding give faint or no radiographs. This simple test, if verified, does not prove the details of the model, but it does place the mechanism squarely in that class of diffusion-analogue processes to which so many other indications point, and removes it unequivocally from the class of flow phenomena. The result needs verifying since it is quite startlingly improbable to those accustomed to think in terms of flow, and since a legume is not the most unexceptionable plant for the test. Opportunity rather than deliberate choice dictated the experimental system. With a nodulated legume, the possibility is not ruled out that ^{14}C-assimilate moved by the phloem to the roots, was transformed in nodules to ^{14}C amino acids, and returned by the xylem to the leaves where it was picked up by the aphids. Such a circulation of label was shown by Pate (1962) in pea. This interpretation of our results seems unlikely on the two grounds: (1) that the aphids are phloem feeders; and (2) that the shoot system does not show the general labelling to be expected from a spread of ^{14}C in the transpiration stream.

This is a special case of the general bi-directional nature of the model. Just as the diverse substances are transported each down its own gradient as discussed in section 3, so if these gradients are in opposite directions, the one sieve tube would transport the substances simultaneously in opposite directions. The conclusion of Chapter 6 was that such movement, while still not proved beyond doubt, was more likely than not on the published evidence.

There are, however, difficulties about this simple view of the spread of labelled assimilate; the experimental facts often show more complex patterns than a steady even spread of label throughout the phloem. For example ^{14}C-assimilate from a leaf near the middle of a stem, on reaching the stem, may travel much faster in one direction, basally or apically, than in the other, whereas the model suggests an equal spread in both directions. Some of the inequality, when the spread is faster basally, could be a result of the leaf traces descending one or more internodes before connecting with the vascular bundles of the stem, making the acropetal path from the leaf looped and long to the internode above its insertion. Label traversing this loop would be analysed as moving basally, aggravating the difference. This may be a real effect, but is not the only one operating, for the relative distribution in the two directions may change with time. In the experiments of Hale and Weaver (1962) the direction of movement of ^{14}C-assimilate was followed in grapevines from leaves of different ages. From the young leaf the label moved only to the apex; from the two or three next older leaves, in both directions; and from older leaves, only to the base of the stem. The appearance of flower buds and fruits as competing sinks further complicated the pattern. It is difficult to avoid the conclusion that sinks and sources have some directing influence on the movement and that this influence varies with the activity of the sinks, altering the pattern. 'Shoot tips and parent vines were more powerful sinks than the cluster during flower development, but not during fruit set.' Here, as always, it is very difficult to distinguish transport from accumulation, sugar in transit from sugar that has arrived and is becoming part of the sink structure both at the end of the channel and alongside it. This distinction is even more blurred in the series of studies by Khan and Sagar (1966, 1967, 1969a & b) on patterns of assimilate movement in tomato where the plants were commonly distributing label for 24 hours. Phloem is an ephemeral tissue, and changing patterns of distribution with time may result from the differentiation of new sieve tubes with different connections, but the case for a directing effect of sources and sinks is strong. If it is really a directing force on translocation and not a result of changing accumulation patterns, the model does not accommodate it.

A simple contradictory example from my own experience is provided by the behaviour of ^{14}C-assimilate in a linear leaf such as that of wheat when the label is applied to a zone in the middle. The model would predict the label should travel equally fast to the tip and base along the parallel veins. In fact it travels only towards the base. I can offer no simple explanation in terms of the model.

5. THE EFFECT OF TEMPERATURE

Changing the temperature of the sieve tubes would influence the model in two ways, through the effects on strand speed (v) and on strand permeability (γ). While the strands were pictured as streaming protoplasm it was easier to correlate their response to temperature with the measured effects on other streaming systems. Now that they must be pictured as much smaller plasma filaments the parallel is more distant, but the effects are possibly similar. Temperatures from 0 to 10 °C are likely to slow their speeds to a fraction of that observed at 20 °C, by the reduced supply of metabolic energy, and by increasing viscosity. Increasing temperatures in the range 20 to 40 °C are likely to increase the speed of the strands considerably, while at some temperature around 50 °C it can be expected that the special protein organisation on which they depend should be de-natur-ed and the speed would fall suddenly to zero.

Those measures of translocation which rely upon the speed component, dye and tracer movements, should show these effects to some extent, and especially at low temperatures. It was seen in Chapter 7 that this is so. The reason that the speed component measures would not entirely reflect

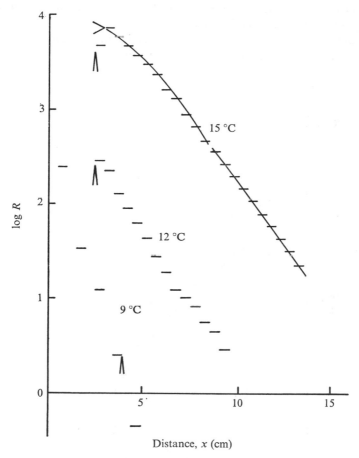

Fig. 16.4. Profiles of ^{14}C developed in cotton petioles at three temperatures in 70 minutes. The junction of lamina and petiole is marked $\wedge$. Distance is measured from the spot on the leaf to which $^{14}CO_2$ was fed. The 9 °C samples were of 1-cm pieces, and have been reduced to the same $\frac{1}{2}$-cm basis as the other two samples. An error curve has been fitted to the top of the 15°C profile with the R_0 value indicated $>$, and with $K = 0.95 \times 10^{-2}$ cm^2 sec^{-1}.

strand-speed responses to temperature, especially at high temperatures, is that they would become complicated by the other effect of temperature: that on strand permeability.

Rising temperature must be expected to increase the permeability of the strand–sap boundary (γ) continuously from 0 °C to the same upper limit of around 50 °C that would stop the movement. At, and even perhaps below this limit the value of γ would increase to that of free diffusion. So even if the strands which carry a speed-measuring substance are moving very fast, they are also becoming more leaky, and the forward progress of the substance is limited by this.

Both effects will operate together to determine the effect of temperature on a measure of translocation based on quantity movement, like the specif-

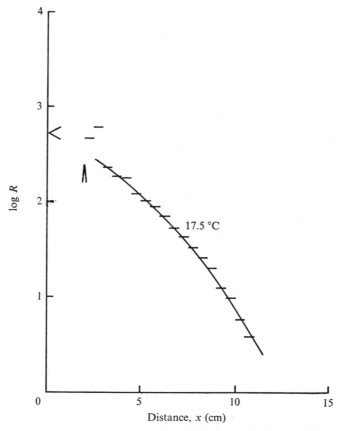

Fig. 16.5. Profile of ^{14}C developed in cotton petiole, as in Fig. 16.4, but at 17.5°C, and in 60 minutes. The fitting constant of the curve is $K = 2.3 \times 10^{-3}$ cm^2 sec^{-1}.

ic mass transfer. They will operate through their effects on the translocation coefficient $K(=rv^2/4\gamma)$. At temperatures just above freezing γ will be
high and v low, but since K depends on v^2, K will be very small and mass
movement slight. As the temperature is raised through 10–20 °C v and γ
will both increase, but it is likely that the squared term will predominate
and K will rise leading to larger movements of mass. Though v may increase linearly with temperature, and v^2 remain a significant part of the
value of K, it seems, from the temperature optima of around 25 °C for
mass-movement measures, that γ increases more rapidly than v^2 in the
range above 20 °C, and comes to dominate the value of K, until by
35–40 °C the strands have become so permeable that they are ceasing to
accelerate diffusion.

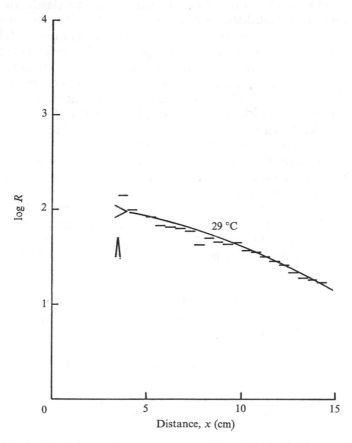

Fig. 16.6. Profile of ^{14}C developed in cotton petiole as in Fig. 16.4, but at 29°C and
in 60 minutes. The fitting constant of the curve is $K = 7.6 \times 10^{-3}$ cm^2 sec^{-1}.

If this rough analysis of the response of the model to temperature is correct three predictions can be made which are open to experimental test: (1) the error curve profile should develop more slowly at lower temperatures, and first in the upper part of the profile, while the forward part of the profile should retain longer a 'loading' shape (see section 8 below); (2) the values of K derived from error profiles formed at different temperatures should show the same temperature response and 25 °C optimum as the mass measures of Chapter 7; (3) at temperatures above 25 °C the error profiles should develop very fast, becoming higher and flatter very quickly, though possibly in a shorter distance.

Some first experiments that support these predictions are given by Whittle (1964a & b) for bracken, and some hitherto unpublished are presented in Figs. 16.4 to 16.6. Cotton plants grown in a glasshouse were transferred to controlled-temperature rooms maintained at temperatures in the range 9 to 29 °C. After an hour or so of equilibration $^{14}CO_2$ was photosynthesised at a spot on a leaf. The profiles of ^{14}C-assimilate developed in the petioles during the next 60 or 70 minutes were found by the assay of $\frac{1}{2}$-cm pieces of the petiole. In all but the 9 °C leaf the profiles are confined to the petiole and start on the distance axis at the appropriate distance from the fed spot of the lamina. In the 9 °C leaf, with a much slower profile, the curve is in the midrib and the top of the petiole and starts from the application site. Commonly the $\frac{1}{2}$-cm section at the junction of the petiole with the lamina was strongly curved and swollen and was difficult to cut precisely, so that the value of the radioactivity per centimetre in this section is often unreliable.

At 9 and 12 °C (in Fig. 16.4) the profile is steep and may well be taken for a straight line. At 15 °C this straight part of the profile survives at the forward end, but the top is curved and can be satisfactorily fitted with an error curve. The straight line points to leakage or, in this context, reflects

TABLE 16.2   *Variation of the translocation coefficient with temperature in cotton petioles* (Data of Fig. 16.4, 16.5, 16.6)

Temperature °C	Translocation coefficient (K) $cm^2\ sec^{-1}$
15	0.95×10^{-3}
17.5	$2.3\ \times 10^{-3}$
29	$7.6\ \times 10^{-3}$

the loading process. All three of these profiles developed in 70 minutes, and with increasing temperature there is evidence of both a greater speed component and a higher permeability (γ). The relative heights of the profiles may mean little, depending as they must on the amount of photosynthesis and loading, both of which will change with temperature, but they fall in a reasonable sequence. In both Fig. 16.5 and 16.6 the translocation time is only 60 minutes. In this shorter time and at 17.5 °C the curved nature of the profile is more pronounced than at 15 °C, and no sign of the steeper forward part survives. The model would interpret this as a marked increase in γ between 15 and 17.5 °C. An error curve has been fitted. At the considerably higher temperature of 29 °C (Fig. 16.6) in the same time the development of the error curve is much further advanced: the profile is flatter, its downward curvature much less. In terms of the model, γ has increased considerably. From 15 to 17.5 °C and even more from there to 29 °C, the speed component v cannot be gauged as accurately as in the 9–15 °C range because of the obscured forward shape of the profile.

The fitted error curves at the three higher temperatures yield values of the translocation coefficient shown in Table 16.2. These clearly increase strongly with temperature in this range, but there is not sufficient data to assess the validity of the predicted optimum.

It will be recalled that Mason and Maskell's value for K was 7×10^{-2} cm^2 sec^{-1} for stem translocation in cotton growing in Trinidad. One must expect varietal differences, and probably also that plants growing habitually at high temperatures have evolved appropriate adjustments. There is a nice point in relation to the temperature of these measurements. With such large variations of translocation with temperature as later work has revealed, the very close relation of rate with gradient shown in Fig. 3.2 (derived from many experiments) becomes remarkable indeed, until we realise how fortunate these experimenters were, and that they partly appreciated the fact:

Whatever the nature of the mechanism may be that is responsible for accelerating diffusion in the sieve-tube, it is surprising that the variation in the activity of this mechanism from one group of plants to another of quite different history and under different conditions should be relatively so small. The environmental conditions varied, however, very little from experiment to experiment. Mean air temperature, in particular, was remarkably uniform, ranging only from 26.1° up to 27.1 °C. (Mason and Maskell, 1928b p. 630.)

They were living in a controlled environment, at the optimum temperature for translocation.

The measurements of K made on bracken at different temperatures (Whittle, 1964b) yielded linear plots of log K v. temperature leading to a

Q_{10} of 2.8–2.9 over the range 13–29 °C. The experimental plants had different ranges of K according as they were grown in the field or in a glasshouse but both populations gave the same Q_{10}. There is no sign of a decrease of K at higher temperatures, but then there are only two determinations above 22 °C, at 24 and 29 °C. As was suggested at the beginning of Chapter 7, the useful measure of translocation on which to assess effects of temperature seems indeed to be the translocation coefficient.

6. EXUDATES, DAMAGE AND PLUGGING

The sap that exudes from a cut aphid stylet has a special interest both from its demonstrated close relationship to the transport in the phloem, and from its long-continuing flow which suggests that the machinery of the sieve tubes is still functioning. From the viewpoint of the model this sap would be the stationary reservoir phase of the sieve tube, initially squeezed out through the food canal by the release of turgor pressure on puncture. The fall in sugar content at the punctured sieve element would make it the lowest point on the sugar gradients, and the strands would at once carry sugar to this element from both directions, drawing on sugar sources of the stem storage and the leaves in proportions depending on distance and the relative ease of the pathways. The rate of sugar output from the stylet in this fashion was calculated on page 112 as equivalent to a specific mass transfer of 2.5 g hr^{-1} cm^{-2} phloem, but this was using the old one-fifth proportion of sieve elements. Using the preferred two-thirds fraction, this would be increased to 8 g hr^{-1} cm^{-2} phloem and would require a local gradient of 10 M sucrose per metre or a drop from 34 per cent to 0 per cent over a decimetre. This is high, but not impossibly so for what is to the sieve tube a crisis situation. The water accompanying this sugar would be drawn osmotically from the xylem sideways along the several centimetres of contributory length, according to the water potential balance at each point, as there is good evidence that it is. The net result will be an outflow of dilute sugar solution (0.5 per cent in Canny, 1961, 5 + per cent in Weatherley, Peel and Hill, 1959) whose sugar content is a large overestimate of the traffic before puncture, whose water content depends on the xylem potential, and whose content of tracer sugar (if such is about) will be related only to gradients of tracer. A predicted consequence would be that the initial drop of sap that appears on cutting should have a higher concentration of sugar than the later samples. This is very difficult to check because the initial drop is so small (20 μm radius, Plate 4) that its separate determination has not been attempted. On the grosser scale of cuts through many sieve tubes, it was seen in Chapter 8 that such a drop in concentration does occur.

The water flow of about 1.0 μl per hour that continues steadily after the initial cut was treated by Weatherley *et al.* (1959) as though it all came in through the surface of the punctured sieve element (170 μm × 23 μm, surface area 13,109 μm²), and they calculate from this a permeability constant for a 5-atmosphere difference in water potential between it and the surrounding tissues,

$$P_w = \frac{10^9}{(13,109 \times 5 \times 60)} = 254 \ \mu\text{m atm}^{-1} \text{ min}^{-1}.$$

They compare this with a usual coefficient of 0.2 μm atm^{-1} min^{-1}. The hypothesis would demand some such high permeability of the sieve element to water. There is no information to balance the probabilities, but it may be recalled that electron images of sieve elements show apparently no tonoplast, the protoplasm in a strange state, and the plasmalemma not easily distinguishable.

Puncture by an aphid stylet (severed) constitutes a high-resistance bleed tapped into the high pressure contents of the sieve element. (With the aphid still attached the resistance is even higher, but the flow measurements are for the severed stylet.) As the elastic tissues of the phloem, under compression from their turgor, contract on this small relief of pressure, the drop is forced out whose volume is the size of the contraction. This small volume change, a drop of 20 μm radius, is

$$\tfrac{4}{3}\pi \left(0.02\right)^3 = 0.0034 \ \text{mm}^3.$$

This may be compared with the volume of our standard willow sieve element which will be

$$0.17\pi \left(0.0115\right)^2 = 0.007 \ \text{mm}^3.$$

So the initial drop is about half the volume of the punctured sieve element, a very reasonable surge. With the aphid present it will be even less. Such a small change of volume is apparently not enough to disrupt the delicate organisation of the sieve tube to such an extent that it ceases to carry sugar. In terms of the model, the plasma-filament strands continue to move and to exchange solutes with the sap.

If, however, the bleed into the high-pressure sieve tubes is larger, as in the case of knife cuts into the phloem, the volume surge is very much greater; large drops appear; the internal organisation is torn apart by the surge, the semi-permeability of the elements is lost, and no further sugar or water is transported to the cut. In those sieve tubes with wide enough sieve pores the surge produces volume flow of contents through a series of many elements, draining the tubes far back from the cut of both sap and some phloem protein, producing the aggregates on the sieve plates known as slime plugs as the flow is filtered through the pores, and a high protein content in the exudate. When the compression forces are fully released, the flow ceases. How far the damage extends back from the cut will depend on the resistance offered by the sieve plates to the flow, but it is known to be many centimetres in some species. Crafts (1939a) showed the displacement of the contents of 250 cm of sieve tubes in cucurbits. This far-reaching damage must be the reason that isolated pieces of phloem do not translocate.

With their main food-transport system so sensitive to damage by pressure release there is a clear selective advantage to land plants, which have been severed by grazing animals and broken by winds since the Devonian, of a sealing system that will reduce the pressure surge, cut down the outflow of sap, and limit the zone of damaged pipeline. This seems to be a main function of the callose. It forms in the sieve pores in response to injury, presumably the pressure release, within a few seconds, constricting the pore. As has been pointed out in Chapter 9, constricting a pore to a fifth of its former diameter reduces the volume flow through it 600 times. With such a response the zone of damaged pipeline will be restricted to the close neighbourhood of the cut.

It is possible that this is the cardinal importance of the sieve plate, whose presence at such frequent intervals across a transport pathway is so difficult to explain. One of the attractions of the model when it used 1-μm strands was that it provided a job for the sieve plates to do: to support the strands as they streamed in the sap, to hold them apart and stop them coalescing. Now that the size scale must be reduced to that of the plasma filaments, this function seems much less important, and the model begins to share with the turgor-driven mass flow hypothesis the paradox of anti-functional components. But if the importance of the plates is to provide medium-sized holes which can be quickly reduced in size with a callose collar, the logical difficulty recedes somewhat.

7. ENERGY REQUIREMENT

There is little useful that can be said about the energy necessary to drive the accelerated diffusion model. We know too little about protoplasmic motions in general and about the filamentous structures that seem to lie at the heart of them (actin, and the filament cores of streaming plant protoplasm observed by Nagai and Rebhun (1966), O'Brien and Thimann (1966), O'Brien and McCully (1970)). The kind of motions that the model would require of filaments of phloem protein, or aggregates of them, have their closest resemblance in the presumed action of these other filaments, which are thought to share with actin some special mechanico-chemical means of turning nucleoside triphosphate bond energy into motion. Without much more detailed understanding of the forces, dimensions and resistances at work in these motions, no calculations can be made to find the work done or the energy expended. Equally, there is no reason to believe that the available 0.5 cal per cc of phloem per hour is insufficient, and Bieleski's work shows that it is translated into high ATP and UTP levels in the sieve-tube lumen.

8. PROFILES AND SPEED

Detailed explicit predictions can be made about the profile of tracer concentration with distance that would be produced as the model operated to carry the tracer from an application site. These predictions are found to agree with many of the real profiles illustrated in Chapter 13, and also suggest other experimental tests. The detailed derivation of these predictions, with the assumptions, limits of confidence and variations of geometry, are contained in Canny and Phillips (1963). A summary of the salient results will be presented here, using the same notation.

Equations 15.1, 15.2 and 15.3 specify the concentrations of sugar in the positive and negative strands and the sap with respect to time and distance. They may be combined into the single equation

$$\frac{\partial^2 S}{\partial t^2} + \frac{4\gamma}{r} \cdot \frac{\partial S}{\partial t} - \frac{v^2 \partial^2 S}{\partial x^2} = 0 \qquad (16.4)$$

where $S = s_1 + s_2$, the sum of the concentrations in the two sets of strands. Investigation of the mass transfer capabilities in section 1 above has already involved the simplified form of this equation which applies in the steady state when the sugar concentration does not vary with time. To study the spread of tracer it is necessary to return the time dependence because initially all the tracer is at the application site, and its concentration increases along the pathway with time. The equation (16.4) is known to engineers as the Telegraph Equation and describes the propagation of a signal along a telegraph wire.

The mode of propagation of tracer loaded into the model depends on the rapidity with which the tracer concentration S changes at the loading site. If S changes sufficiently slowly, the first term of (16.4) is neglected and

$$\frac{\partial S}{\partial t} = \frac{rv^2}{4\gamma} \cdot \frac{\partial^2 S}{\partial x^2}. \qquad (16.5)$$

This is the classical diffusion equation with the value of the diffusion coefficient, the constant $rv^2/4\gamma$. Its appearance here explains both why the strand model can produce error curves like diffusion (Chapter 13, Model 4), and the identification of equation 16.1 that the translocation coefficient is expressible as $rv^2/4\gamma$. For rapid changes of S with time, the first term of equation 16.4 is larger than the second and

$$\frac{\partial^2 S}{\partial t^2} - \frac{v^2 \partial^2 S}{\partial x^2} = 0 \qquad (16.6)$$

which is the classical equation for the propagation of a wave at speed v. So sudden tracer changes at the loading site will be propagated as wave profiles of unchanging shape and that shape will be determined by the loading process. Their speed will be the strand speed, v.

Now the yardstick by which we may measure whether a change in S is fast or slow, is a constant of the model called the 'strand diffusion time'; its symbol is θ; it is given by

$$\theta = r/2\gamma \qquad (16.7)$$

and is measured in seconds. If S changes appreciably in a time short compared with θ, the change will be propagated as a wave of constant shape but diminishing amplitude, Fig. 16.7. The tracer moving down the positive strands has the same profile as when loaded and forms an attenuating front. Some tracer permeates to the sap, but in the region of the front this is a small fraction. Still less can diffuse to the negative strands in this time.

If changes in S take times long compared with θ, the advancing profile will be a series of error curves as shown in Figs. 13.3 and 13.4 until the tracer reaches a sink at the other end of the channel.

Consider then the changing pattern with time after loading of tracer into the channel where the apparatus of the model has its characteristic value of θ. At times short compared with θ the loading profile is carried along unchanged in shape. This is the stage of the wave profile. After times longer than θ the tracer will appear spread out into error curves. As time goes on the error curve spreads to greater distances, developing through a predictable series of shapes until it reaches a sink. When it does so the concentration increases gradually along the whole length of channel until the curvature of the distribution is lost and the final linear steady-state gradient is reached. Successive error curves would be expected to yield the same values of K if the machinery does not change its nature and functioning along the channel.

Between these two stages, at times comparable with θ, the profile will change from the wave type to the diffusion type by way of transition forms intermediate between the two. The advancing loading profile is reduced

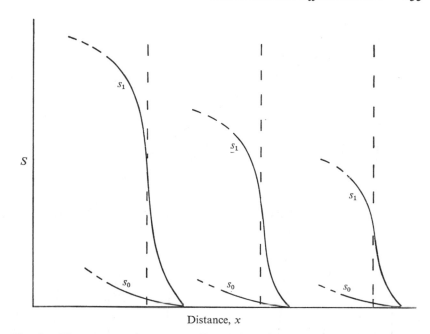

Fig. 16.7. The progress of a wave profile: at times short compared with the strand diffusion time, the pattern of tracer produced by the loading mechanism would be carried forward by the model unchanged in shape, but with slowly diminishing height. Re-drawn from Canny and Phillips (1963), Fig. 5.

in height, but its shape retains the history of loading. Behind the front, the behaviour is diffusive, and the top of the profile develops into an error curve. This is illustrated in Fig. 16.8.

The speed of the strands is identifiable with certainty only during the first two stages while the wave profile remains. Once it has vanished and been replaced by the diffusion profiles, the advance of tracer no longer keeps up with the advance of the strands.

Let us now turn back to the representative collection of experimental profiles, and, armed with the suggestions arising from these predictions, consider again the shapes that appeared.

Fig. 13.8: ^{32}P-phosphate (Biddulph and Cory). This crude data is sufficiently explained by leakage as already discussed.

Fig. 13.9: ^{14}C-assimilate (Vernon and Aronoff). The fit of the error curve is good over the whole range, suggesting that the strand diffusion time in these *Soya* plants was small compared with 15 minutes.

Fig. 13.10: ^{14}C-assimilate (Vernon and Aronoff). The contrast with

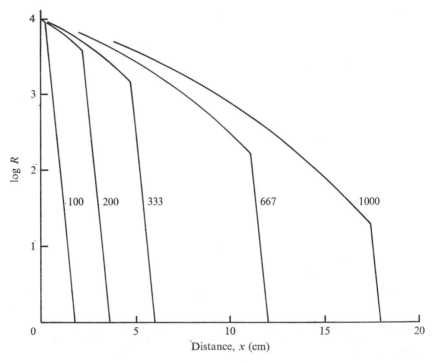

Fig. 16.8. Theoretically constructed transition profiles developed by the accelerated-diffusion model at successive translocation times. The strand diffusion time was chosen as 100 sec and the strand speed as 1.8×10^{-2} cm sec^{-1}, giving a value for K of 1.6×10^{-2} cm^2 sec^{-1}. A loading profile has been chosen arbitrarily with a straight-line slope on the semi-log plot. It proceeds along the distance axis at the strand speed, remaining the same slope, but getting smaller. At its upper end the error curves develop as shown. A constant value of R_0 has been assumed at $x = 0$. In practice R_0 is likely to increase with time as more tracer is loaded.

Fig. 13.9 when the pathway is not darkened has already been discussed in terms of the re-assimilation that simulates leakage.

Fig. 13.11: ^{14}*C-assimilate (Biddulph and Cory).* Rudimentary data not showing any clear shape, but suggesting a straight line again. Possibly re-assimilation.

Fig. 13.12: ^{14}*C-assimilate (Gage and Aronoff).* Same remarks as for 13.9; another strand diffusion time for *Soya* less than 20 minutes.

Fig. 13.13: THO-assimilate (Gage and Aronoff). As for Fig. 13.12.

Figs. 13.14 and 13.15: ^{14}*C-assimilate (Mortimer).* These are two replicates from *Beta* petioles after 30 minutes. The data may be interpreted in three ways: (*a*) the points do not lie on an error curve at all; (*b*) they are a bit ragged and might lie on an error curve between the two limits shown in Fig. 13.14; or (*c*) they are transition profiles with an error curve developing at the top, and a steeper loading profile preserved at the forward end. If the last is right, then the strand diffusion time is not very small compared with 30 minutes.

Fig. 13.16: [14]*C-assimilate (Mortimer).* Another profile from *Beta* petiole after 20 minutes in which the error curve fits snugly over 200-fold change of R, suggesting that the strand diffusion time is in fact short compared with 20 minutes.

Fig. 13.17: [14]*C-assimilate (Mortimer).* Only one of these replicates from the 15-minute samples is fitted by the error curve. The other is erratic. Again there is no sign of a surviving wave profile.

Fig. 13.18: [14]*C-assimilate (Lawton).* This data from *Dioscorea* shows a feature quite commonly seen in real profiles: a good error curve on the forward end, and an R_0 value well down the distance axis, leaving an even plateau behind it where the radioactivity is constant with distance. This suggests that the exchange space into which the label is equilibrating is filled in this part of the plant. The translocation time of 130 minutes is long compared with any expected strand diffusion time. It is not surprising that no loading form survives.

Fig. 13.19: [14]*C-assimilate (Lawton).* The times these *Dioscorea* plants have been translocating (4, 5 and 6 hours) must be very long compared with the strand diffusion time. The variation of K cannot be certainly ascribed to any one of the three variables: time, distance or between plants. Broadly it may be said that the curves show the expected progress with distance and upward flattening with time.

Fig. 13.20: [14]*C-assimilate (Fisher).* More *Soya* data showing pure error curves and pointing to a strand diffusion time shorter than the earliest sample at 10 min. The beautiful agreement in values of K throughout the time and between plants has already been pointed out. The value of R_0 increases rapidly (× 10) between 10 and 17 minutes, and again between 17 and 25 minutes, but not thereafter, suggesting a time scale during which the peak concentration in the sieve tubes is achieved by the loading mechanism.

Fig. 13.21: [11]*C-assimilate (Moorby* et al.*).* It will be recalled that these important profiles are all from a single plant of *Soya*. They show strikingly the kind of behaviour predicted for transition profiles. The steep forward ends would represent the loading process. If this is so, then the strand diffusion time of this *Soya* plant must be much longer than those just discussed in Figs. 13.9, 13.10, 13.12, 13.13 and 13.20. The values of K derived from the first three curves, from 17 to 23 min, are in good agreement, but after this K has nearly doubled by 32 min. This would imply changing efficiency of the model in different parts of the plant.

Fig. 13.22 (= 16.5) and the associated Figs. 16.4, 16.6: ^{14}C-assimilate (original). This set of curves from cotton has already been discussed in some detail in relation to the effects of temperature in section 5. It may now be better appreciated that the straight-line part of the curve at 15 °C can be interpreted as a loading profile, and that the 12 °C and 9 °C curves would on the same view be pure loading profiles, proceeding in the wave form. It is clear that the strand diffusion time will increase with rising temperature and from these figures we may say that

$$
\begin{aligned}
&\text{at} \quad 9\,°C\ \theta > 70\ \text{min}\\
&\text{at} \quad 12\,°C\ \theta \simeq 70\ \text{min}\\
&\text{at} \quad 17.5\,°C\ \theta < 60\ \text{min}\\
&\text{at} \quad 29\,°C\ \theta \ll 60\ \text{min.}
\end{aligned}
$$

Fig. 13.23: ^{14}C-assimilate (original). This curve from sunflower shows even more clearly than Fig. 13.18 the steady level of label behind the advancing error curve. The translocation time is much longer than the strand diffusion time.

Figs. 13.24a and b: ^{14}C-assimilate (original). These profiles in strawberry petioles again could be objected to as error curves since the curve fits over only two to three powers of ten, and part of the source end is flattening out. The curves have been chosen to fit the forward end rather than the upper part because these long times of 70 and 90 minutes must be longer than the strand diffusion time, and it seems unlikely that the forward part retains any qualities related to loading.

Fig. 13.25: ^{14}C-assimilate (Hartt et al.). In these profiles in the stem of sugar cane the scale of events is large; the distances are great and the times are long. The profiles must be past the transition stage, and look to be dominated by sideways accumulation from the straight-line slopes of the front. Indeed this is most likely in this plant whose stem contains large pools of extra-phloem sucrose which are known to accumulate the sugar actively.

Figs. 13.26, 13.27 and 13.28: ^{14}C-assimilate (Lee et al.). These sunflower curves have resemblances to those of Fig. 13.23 and Fig. 8 of Canny and Phillips (1963). From all these it would seem that the strand diffusion time is rather long in sunflower, and transition curves are often found. However, the time scale seems to me to be altogether too long for the model with reduced strand radii, and there is probably some other explanation for these characteristic shapes.

From this review of observed profiles it is possible to draw the conclusion that investigations so far have been done too clumsily, with translocation times too long, sieve-tube samples too gross, samples too variable. The amount of useful information that can be gleaned from data of this kind is most in the early stages of the spread of tracer, while the wave form survives and can give data on speed from its position and on loading from its shape. The larger-scale experiments just discussed have a limited usefulness, confined really to the determinations of K. They are easy and rapid ways of arriving at a value for this coefficient, and it has been one of my special concerns to argue that the coefficient is worth knowing. Its true importance and usefulness remain to be established by an experiment not yet attempted: the simultaneous measurement of K by the two independent methods, chemical and tracer. If a value of K obtained by repeating a Mason–Maskell type of analysis of the proportionality of rate with gradient were made on plants that were carying tracer and whose profiles yielded an independent estimate of K, the real relevance or otherwise of the diffusion analogy could be ascertained. To learn more than this about translocation from profiles it is going to be necessary to refine the techniques to deal with very small pieces of phloem after translocation times of not many seconds, and the labour of this may not be worth while. Nevertheless it is worth completing the argument by pointing out what might be measured and how, in a general discussion of the parameters of the model.

9. PARAMETERS OF THE ACCELERATED DIFFUSION MODEL

The translocation coefficient K is an experimental reality derivable in two independent ways, and has an existence independent of the model. If the view developed in these pages is a sound one it is the most important measure of the translocating capacity of a piece of phloem, and, coupled with information about the gradient, will lead to predictions of mass transfer. The various values of this coefficient for the translocated products of assimilation, mentioned in different parts of this book, have been collected in Table 16.3, grouped by genera. All except the first are derived from profiles of advancing tracer. Considering the variety of plant, organ and experimental conditions, the range is not large: the extremes of the 29 values are 3.3×10^{-4} and 7×10^{-2}, and most lie between 4×10^{-3} and 4×10^{-2}. They are in the range that would adequately account for the mass transfers of Table 1.1. It should be noted that no measures of K exist for the stalks of developing fruits which account for many of the high values of mass transfer, and that with the exception of Mason and Maskell's cotton plants, all the plants of Table 16.3 are pot-grown seedlings.

These values of K may be interpreted in terms of the model to test

TABLE 16.3. Values of the translocation coefficient for translocated assimilate (K) and estimates of strand speed (v) derived from the experiments discussed, together with calculated values of strand diffusion time (θ)

Plant	Author	Text Figure	K cm^2 sec^{-1}	t sec	v cm sec^{-1}	Remarks	$\theta(=2K/v^2)$ sec
Gossypium	Mason & Maskell (1928a & b)	3.2	7×10^{-2}	–	–	chemical estimate	–
Soya	Vernon & Aronoff (1952)	13.9	4.6×10^{-2}	1200	2.0×10^{-2}	diffusion profile	230
Soya	Gage & Aronoff (1960)	13.12	1.2×10^{-2}	1200	2.1×10^{-2}	diffusion profile	55
Soya	Gage & Aronoff (1960)	13.13	1.7×10^{-2}	900	1.8×10^{-2}	diffusion profile	105
Soya	Fisher (1970)	13.20	6.0×10^{-3}	600	1.3×10^{-2}	diffusion profile	70
Soya	Fisher (1970)	13.20	6.0×10^{-3}	1020	1.5×10^{-2}	diffusion profile	53
Soya	Fisher (1970)	13.20	6.0×10^{-3}	1500	1.6×10^{-2}	diffusion profile	48
Soya	Fisher (1970)	13.20	6.0×10^{-3}	2100	1.6×10^{-2}	diffusion profile	48
Soya	Moorby (unpubl.)	13.21	3.5×10^{-3}	1020	2.9×10^{-2}	transition profile	8
Soya	Moorby (unpubl.)	13.21	3.3×10^{-3}	1200	2.5×10^{-2}	transition profile	11
Soya	Moorby (unpubl.)	13.21	3.8×10^{-3}	1380	2.3×10^{-2}	transition profile	14
Soya	Moorby (unpubl.)	13.21	4.5×10^{-3}	1560	2.2×10^{-2}	transition profile	19
Soya	Moorby (unpubl.)	13.21	5.8×10^{-3}	1740	?use 2.2	transition profile	25
Soya	Moorby (unpubl.)	13.21	6.9×10^{-3}	1920	?use 2.2	transition profile	29
Beta	Mortimer (1965)	13.14	6.0×10^{-2}	1800	1.6×10^{-2}	?transition profile	470
Beta	Mortimer (1965)	13.15	4.3×10^{-2}	1800	1.5×10^{-2}	?transition profile	380
Beta	Mortimer (1965)	13.16	3.2×10^{-2}	1200	2.1×10^{-2}	diffusion profile	148
Beta	Mortimer (1965)	13.17	2.0×10^{-2}	900	1.7×10^{-2}	diffusion profile	138
Dioscorea	Lawton (1967)	13.18	3.5×10^{-2}	7800	1.4×10^{-3}	diffusion profile	35,000
Dioscorea	Lawton (1967)	13.19	2.1×10^{-4}	14,400	1.5×10^{-3}	diffusion profile	187
Dioscorea	Lawton (1967)	13.19	4.0×10^{-3}	18,000	2.4×10^{-3}	diffusion profile	1380
Dioscorea	Lawton (1967)	13.19	1.3×10^{-2}	21,600	4.2×10^{-3}	diffusion profile	1450
Gossypium	Canny	16.14	9.5×10^{-4}	4200	3.1×10^{-3}	?transition, 15°C	196
Gossypium	Canny	13.22	2.3×10^{-3}	3600	3.7×10^{-3}	diffusion, 17.5°C	330
Gossypium	Canny	16.16	7.65×10^{-3}	3600	$>4.2 \times 10^{-3}$	diffusion, 29°C	<900
Helianthus	Canny & Phillips (1963)	–	4.0×10^{-2}	2400	6.7×10^{-3}	transition profile	1800
Helianthus	Canny	13.23	1.8×10^{-3}	8100	3.9×10^{-3}	diffusion profile	240
Fragaria	Canny	13.24a	3.3×10^{-4}	4200	2.4×10^{-3}	diffusion profile	114
Fragaria	Canny	13.24b	8.6×10^{-4}	5400	2.2×10^{-3}	diffusion profile	354

whether the properties of the model that would be required to explain the experimental data lead to possible conclusions. We have seen in equation 16.1 that $K = rv^2/4\gamma$, so that knowing K, v and r, we can arrive at an estimate of γ. It would be very helpful to be able to measure γ with some precision, for it should depend on the substance being translocated, and a knowledge of the relation of γ to the chemical nature of the substance would provide information about the properties and structure of the boundary between strand and sap, just as the permeability of cells to solutes gives information about the cell membrane. But r is not known with any precision. Thaine-type strands had $r = 1$ μm; filaments of P-protein have $r = 125$ Å – a wide range of uncertainty. So to start with it is better to restrict calculation to the ratio of the two, which has been called the strand diffusion time,

$$\theta = r/2\gamma. \tag{16.7}$$

This is readily derivable from experimental curves that yield K, if the speed of the strands (v) can also be estimated. As was said in section 8, this can be done accurately only during the wave and transition stages of the profiles, and most of the data of the table are for the later, diffusion stage. However, the distance reached by the furthest detected tracer gives a minimum estimate of v, worse at later times, and for what they are worth, these have been set down also in the table. The translocation time t is known with some precision, and is recorded for each value of K and v. The 'Remarks' column lists mostly the apparent shape of the profile in the terms of section 8, and other relevant facts. The final column lists the calculated values of $\theta = 2K/v^2$ in seconds. It should be noted that since v is often uncertain and θ depends on v^2, these values will not individually be very trustworthy. The body of them, however, is full of interest. They suggest that the different genera have characteristic ranges of θ; that for *Soya* θ is short, mostly less than a minute; that others have values of θ of the order of 150–300 seconds; and that the longest strand diffusion times are of the order of a thousand or so seconds. The *Dioscorea* value looks like a rogue. Most of those showing the diffusion shape have, as would be expected, values of t large compared with θ. Those listed as transition profiles ought to have values of t not very large compared with θ, and this is so for the sunflower curve and the two sugar beet curves, but not for the cotton, or for Moorby's data on soybean. This would suggest that the interpretation of these latter curves as transition ones is unsound.

Now a comparison can be made with this considerable mass of data on θ, which would be unsound in a individual case: the comparison with strand radius and permeability. In Table 16.4 is assembled a selection from the range of possible values of θ (from 20 to 1000 sec) and of r

TABLE 16.4.   *The consequences in terms of strand perme-ability (γ) of a range of values of radius (r) and strand diffu-sion time θ. Values of γ in cm sec^{-1}.*

r cm	θ sec		
	20	100	1000
10^{-4}(1 μm)	2.5×10^{-6}	5×10^{-7}	5×10^{-8}
10^{-6}(100 Å)	2.5×10^{-8}	5×10^{-9}	5×10^{-10}

(from 10^{-4} to 10^{-6} cm), and for each is calculated the value of γ. It will be apparent that the values of γ for sucrose needed to attain a strand diffusion time of 20 seconds with a 1-μm-radius strand is very large indeed – half that of free diffusion – and would lead to no transport or impossibly high values of v. Looked at in this way, the smaller strand version of the model is much more satisfactory in that with intermediate values of γ around 10^{-9} (which fit well with the experimental values of K, v, τ, etc., see section 1) the observed strand diffusion times are readily attained. So the kinetic data support the histological data in denying the presence of strands visible in the light microscope and requiring strands the size of phloem protein filaments or small aggregates of them.

10. CONCLUSION

Here the argument at present rests. What data there are available support in general terms the hypothesis of the model, and at some points the correspondences seem too striking to be coincidental. Detailed predictions, quantitative and qualitative, derived from the model have proved accurate in a number of instances. Other predictions either remain to be tested (many are simple to test; a few much more difficult), or appear to be at variance with the available information. It has been an important part of my purpose to identify these latter, to bring them and the reasons for their relevance out into view and stimulate work and thought concerning them. It is certain that the model emphasised in this chapter is not the only one that would have the properties necessary to fit the assembled facts. Others can no doubt be devised that would share its salient properties: the analogy with diffusion, the transport capacity, the patterns of spread, the development of profiles. Error functions will be generated by any system with random probability in its motions, even, as we have seen, by flow systems with a strong uni-directional component. Equally it is clear that these are the properties of the translocation system, and any model to explain translocation should have these properties. If all this experimenting and reasoning has any single conclusion, it is to emphasise the comparison that Mason and Maskell drew: 'Thus sugars appear to move in the sieve-tube under unit gradient at about the same rate as they would diffuse in air if they existed in the gaseous state.' One feasible way in which this could happen has been highlighted. There are surely others; and these could depend on as yet unsuspected properties of protoplasm as it exists in these unique cells.

Fancy has rambled far from the pumpkin with which we began, and has found food of sufficient variety to satisfy many palates, and perhaps to upset a few digestions. If the pumpkin has not yet become a coach to carry us triumphant, proclaiming a proven doctrine, this is not for want of diligent effort by generations of ingenious and thoughtful men patiently whittling away at the problem. And if it be objected that so far their labours have produced from the pumpkin only a hollow shell, it can be answered that the flickering light of the All Hallows' candle gleams through the eye-slits of a discernible face.

Appendix 1

Fitting error functions to data and calculating translocation coefficients

The equation (13.1) of the diffusive spread of radioactivity is:

$$\frac{R_x}{R_0} = 1 - \text{erf} \frac{x}{2 (Kt)^{\frac{1}{2}}} \qquad (\text{A1.1})$$

where R_x is the radioactivity at any point x,

R_0 is the radioactivity at the place where x is measured from,

x is the distance in centimetres,

t is time in seconds,

K is the diffusion (translocation) coefficient in $cm^2 \ sec^{-1}$.

R_x and R_0 can be in any units as long as they are the same. Take Vernon and Aronoff's data of Fig. 13.9 plotted as $\log R_x$ v. x. We first fix on a value of R_0, the radioactivity at the point we start measuring distance from. Distance does not have to be zero at the source leaf, and will not be if the profile has moved some distance through the plant leaving a uniform concentration of tracer behind it (c.f. Fig. 13.28). In Fig. A1.1 we may choose R_0 at $x=0$, and extrapolating the line of points back to meet the ordinate, choose $\log R_0 = 4.96$. We are now interested in the distances at which the radioactivity falls to various fractions of R_0, i.e. where

$$\frac{R_x}{R_0} = 1/10, \ 1/100, \ 1/1000, \ \text{etc.}$$

These particular fractions will be very easy to find on the plot of $\log_{10} R_x$. A rough line is sketched through the points (dotted), and the places marked where R_x/R_0 has the above values, i.e. at $\log R = 3.96, 2.96, 1.96$. If we need intermediate points then $1/3$, $1/30$, $1/300$, are convenient and easily found by counting down 0.477 units of the $\log R$ axis.

The equation says that

$$\text{erf} \left[x/2 (Kt)^{\frac{1}{2}} \right] = 1 - R_x/R_0.$$

We shall be interested in just the few error functions that correspond with R/R_0 of the above sizes. Take $R_x/R_0 = 1/10$. Therefore

$$\text{erf} \left[x/2 (Kt)^{\frac{1}{2}} \right] = 1 - 1/10 = 0.900.$$

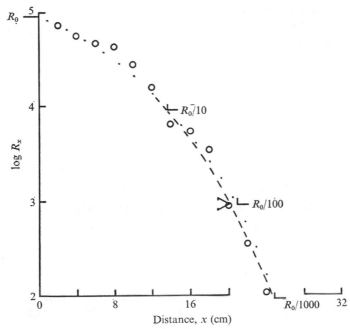

Fig. AI.I. The data of Fig. 13.9 (Vernon and Aronoff) used to illustrate the fitting of an error curve. The experimental points are shown, $\bigcirc$; a rough curve drawn by eye through the points, ...; the points where this curve crosses $R_0/10$, $R_0/100$ and $R_0/1000$, $\llcorner$; the corrected position of the $R_0/100$ point, $>$; and the chosen error curve where it is not co-incident with the dots is shown, ----.

TABLE AI.I   *Numbers whose error function is $(1 - R_x/R_0)$ for simple values of R_x/R_0*

R_x/R_0	$x/2(Kt)^{\frac{1}{2}}$
4/5	0.180
2/3	0.304
1/2	0.48
1/3	0.68
1/5	0.909
1/10	1.16
1/30	1.51
1/100	1.82
1/300	2.075
1/1000	2.33
1/10000	2.75

Turn to Appendix 2 and see that the number whose error function is 0.900 is 1.16. Therefore the value of $x/2(Kt)^{\frac{1}{2}}$ is 1.16. A small table (AI.1) will contain all the values needed for almost any data.

Return to Fig. AI.1 and draw up a table for the calculation:

t sec	$\log R_0$	R_x/R_0	x	$x/2(Kt)^{\frac{1}{2}}$	$2(Kt)^{\frac{1}{2}}$
1200	4.96	1/10		1.16	

We find our dotted curve cuts $\log R_x = 3.96$ at $x = 12.8$, so we place that in the table and divide it by 1.16 to get the value of $2(Kt)^{\frac{1}{2}}$:

t	$\log R_0$	R_x/R_0	x	$x/2(Kt)^{\frac{1}{2}}$	$2(Kt)^{\frac{1}{2}}$
1200	4.96	1/10	12.8	1.16	11.0

Now do the same for $R_x/R_0 = 1/100$, and get a second estimate of $2(Kt)^{\frac{1}{2}}$:

1200	4.96	1/100	20.45	1.82	12.3

and for $R_x/R_0 = 1/1000$:

1200	4.96	1/1000	24.4	2.33	11.0

Choose an average, or experiment with various values of $2(Kt)^{\frac{1}{2}}$ to get the line that fits closest. Re-calculate the x values for this, say 11.0:

R_x/R_0	$x/2(Kt)^{\frac{1}{2}}$	x
1/2	0.48	5.6
1/10	1.16	12.8
1/100	1.82	20
1/1000	2.33	44.4

Then using the chosen value of $2(Kt)^{\frac{1}{2}}$, find K:

$$2(Kt)^{\frac{1}{2}} = 11.0$$
$$\therefore (Kt)^{\frac{1}{2}} = 5.5$$
$$\therefore Kt = 30$$
$$\therefore K = 30/1200 = 4.6 \times 10^{-2} \text{ cm}^2 \text{ sec}^{-1}.$$

N.B. Because of the squaring of (Kt) the value of K you get will be very sensitive to variations of $(Kt)^{\frac{1}{2}}$. The data must lie closely on the smooth curve.

So the two values of x corresponding to our selected $2(Kt)^{\frac{1}{2}}$, and $R_x = R_0/10$ and $R_x = R_0/1000$, both lie on the error curve, but the dotted curve lies too far to the right at $R_x = R_0/100$. So returning to this entry in the table, the value of x necessary to give $2(Kt)^{\frac{1}{2}} = 11.0$ is $1.82 \times 11.0 = 20.02$. The point is shifted back to this value of x (arrow) and now the error curve lies through all four points. It is coincident with the dots in the upper part, and shown dashed from where it diverges.

Appendix 2

$$\mathrm{Erf}(q) = \frac{2}{\sqrt{\pi}} \int_{o}^{q} e^{-y^2}\, dy$$

q	I	q	I	q	I	q	I	q	I	q	I
.01	.01128	.51	.52924	1.01	.84681	1.51	.96728	2.01	.99552	2.51	.99961
.02	.02256	.52	.53790	1.02	.85084	1.52	.96841	2.02	.99572	2.52	.99963
.03	.03384	.53	.54646	1.03	.85478	1.53	.96952	2.03	.99591	2.53	.99965
.04	.04511	.54	.55494	1.04	.85865	1.54	.97059	2.04	.99609	2.54	.99967
.05	.05637	.55	.56332	1.05	.86244	1.55	.97162	2.05	.99626	2.55	.99969
.06	.06762	.56	.57162	1.06	.86614	1.56	.97263	2.06	.99642	2.56	.99971
.07	.07886	.57	.57982	1.07	.86977	1.57	.97360	2.07	.99658	2.57	.99972
.08	.09008	.58	.58792	1.08	.87333	1.58	.97455	2.08	.99673	2.58	.99974
.09	.10128	.59	.59594	1.09	.87680	1.59	.97546	2.09	.99688	2.59	.99975
.10	.11246	.60	.60386	1.10	.88021	1.60	.97635	2.10	.99702	2.60	.99976
.11	.12362	.61	.61168	1.11	.88353	1.61	.97721	2.11	.99715	2.61	.99978
.12	.13476	.62	.61941	1.12	.88679	1.62	.97804	2.12	.99728	2.62	.99979
.13	.14587	.63	.62705	1.13	.88997	1.63	.97884	2.13	.99741	2.63	.99980
.14	.15695	.64	.63459	1.14	.89308	1.64	.97962	2.14	.99753	2.64	.99981
.15	.16800	.65	.64203	1.15	.89612	1.65	.98038	2.15	.99764	2.65	.99982
.16	.17901	.66	.64983	1.16	.89910	1.66	.98110	2.16	.99775	2.66	.99983
.17	.18999	.67	.65663	1.17	.90200	1.67	.98181	2.17	.99785	2.67	.99984
.18	.20093	.68	.66378	1.18	.90484	1.68	.98249	2.18	.99795	2.68	.99985
.19	.21184	.69	.67084	1.19	.90761	1.69	.98315	2.19	.99805	2.69	.99986
.20	.22270	.70	.67780	1.20	.91031	1.70	.98379	2.20	.99814	2.70	.99987
.21	.23352	.71	.68467	1.21	.91296	1.71	.98441	2.21	.99822	2.71	.99987
.22	.24430	.72	.69143	1.22	.91553	1.72	.98500	2.22	.99831	2.72	.99988
.23	.25502	.73	.69810	1.23	.91805	1.73	.98558	2.23	.99839	2.73	.99989
.24	.26570	.74	.70468	1.24	.92051	1.74	.98613	2.24	.99846	2.74	.99989
.25	.27633	.75	.71116	1.25	.92290	1.75	.98667	2.25	.99854	2.75	.99990
.26	.28690	.76	.71754	1.26	.92524	1.76	.98719	2.26	.99861	2.76	.99991
.27	.29742	.77	.72382	1.27	.92751	1.77	.98769	2.27	.99867	2.77	.99991
.28	.30788	.78	.73001	1.28	.92973	1.78	.98817	2.28	.99874	2.78	.99992
.29	.31828	.79	.73610	1.29	.93190	1.79	.98864	2.29	.99880	2.79	.99992
.30	.32863	.80	.74210	1.30	.93401	1.80	.98909	2.30	.99886	2.80	.99992
.31	.33891	.81	.74800	1.31	.93606	1.81	.98952	2.31	.99891	2.81	.99993
.32	.34913	.82	.75381	1.32	.93807	1.82	.98994	2.32	.99897	2.82	.99993
.33	.35928	.83	.75952	1.33	.94002	1.83	.99035	2.33	.99902	2.83	.99994
.34	.36936	.84	.76514	1.34	.94191	1.84	.99074	2.34	.99906	2.84	.99994
.35	.37938	.85	.77067	1.35	.94376	1.85	.99111	2.35	.99911	2.85	.99994
.36	.38933	.86	.77610	1.36	.94556	1.86	.99147	2.36	.99915	2.86	.99995
.37	.39921	.87	.78144	1.37	.94731	1.87	.99182	2.37	.99920	2.87	.99995
.38	.40901	.88	.78669	1.38	.94902	1.88	.99216	2.38	.99924	2.88	.99995
.39	.41874	.89	.79184	1.39	.95067	1.89	.99248	2.39	.99928	2.89	.99996
.40	.42839	.90	.79691	1.40	.95229	1.90	.99279	2.40	.99931	2.90	.99996
.41	.43797	.91	.80188	1.41	.95385	1.91	.99309	2.41	.99935	2.91	.99996
.42	.44747	.92	.80677	1.42	.95538	1.92	.99338	2.42	.99938	2.92	.99996
.43	.45689	.93	.81156	1.43	.95686	1.93	.99366	2.43	.99941	2.93	.99997
.44	.46623	.94	.81627	1.44	.95830	1.94	.99392	2.44	.99944	2.94	.99997
.45	.47548	.95	.82089	1.45	.95970	1.95	.99418	2.45	.99947	2.95	.99997
.46	.48466	.96	.82542	1.46	.96105	1.96	.99443	2.46	.99950	2.96	.99997
.47	.49375	.97	.82987	1.47	.96237	1.97	.99466	2.47	.99952	2.97	.99997
.48	.50275	.89	.83423	1.48	.96365	1.98	.99489	2.48	.99955	2.98	.99997
.49	.51167	.99	.83851	1.49	.96490	1.99	.99511	2.49	.99957	2.99	.99998
.50	.52050	1.00	.84270	1.50	.96611	2.00	.99532	2.50	.99959	3.00	.99998
										∞	1.00000

Appendix 3
Symbols used: definitions and units

C	concentration of solute	g cm^{-3}
D	coefficient of molecular diffusion	cm^2 sec^{-1}
x	distance	cm
A	surface area	cm^2
K	translocation coefficient	cm^2 sec^{-1}
P	pressure	dyne cm^{-2}
	or pressure gradient	atmosphere cm^{-1}
l	length of tube, etc.	cm
v	speed	cm sec^{-1}, etc.
η	viscosity	poise
r	radius of strand	cm
R	radioactive concentration	μCi, c.p.m., etc. per cm or, sometimes, per cm^3
R_0	the maximum value of R, taken as the distance origin of a profile; the value of R at $x=0$	μCi cm^{-1}, etc.
R_x	the value of R at a particular distance x	μCi cm^{-1}, etc.
t	time	sec, hr, etc.

$$\text{erf}(q) \quad \text{the error function } \text{erf}(q) = \frac{2}{\sqrt{\pi}} \int_0^q e^{-y^2} \, dy$$

	(Tabulated in Appendix 2)	
a	radius of a pipe	cm
U	mean velocity of fluid in a pipe	cm sec^{-1}
N_R	Reynolds number	
v	dynamic viscosity	stoke
u_0	maximum velocity of fluid in a pipe	cm sec^{-1}
k	apparent coefficient of diffusion of solute in a solution flowing in a pipe	cm^2 sec^{-1}
N_a	number of pipes of radius a in a bundle of pipes of different radii	
n	number of hypothetical strands in a sieve tube in the Canny/Phillips model	

s	concentration of translocated solute, e.g. sucrose	$\mathrm{g\ cm^{-3}}$
s_1	value of s in positive strands	$\mathrm{g\ cm^{-3}}$
s_2	value of s in negative strands	$\mathrm{g\ cm^{-3}}$
s_0	value of s in sap between strands	$\mathrm{g\ cm^{-3}}$
γ	permeability of strand–sap boundary to translocated solute	$\mathrm{cm\ sec^{-1}}$
σ	the difference in concentration between positive and negative strands	$\mathrm{g\ cm^{-3}}$
S	sugar concentration at the origin of the gradient, $x=0$	$\mathrm{g\ cm}^{\ 3}$
T	net transport of solute along a sieve tube	
ρ	ratio of strand cross section to sieve-tube cross section	
$l_{\max}$	the maximum distance over which a specified mass transfer is possible. The point at which s_2 becomes zero	cm
τ	the rate of mass transfer per unit area of sieve tube; specific mass transfer	$\mathrm{g\ sec^{-1}\ cm^{-2}}$ $\mathrm{g\ hr^{-1}\ cm^{-2}}$
θ	strand diffusion time	sec

Appendix 4
A steam generator for killing tissue

Wilbur said it wouldn't fly; Orville said it wouldn't fly.

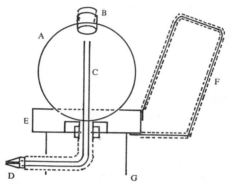

Fig. A4.1.

For the kind of 'ringing' experiment referred to several times in which the living cells of a petiole or stem are killed by steam while the contents of the tracheary elements are left intact, there is needed a small, portable generator. A satisfactory design that I have evolved is offered here. It works well in the field: it is self-contained, provides an indefinite duration of a small, precise jet of steam without spitting, is not upset by wind, nor hampered by supply cables. It is easily constructed in a simple workshop.

The basic features of the design are illustrated in Fig. A4.1. A spherical boiler is made conveniently from a 4-in. diameter copper ballcock (A). It may be filled through a tube (B) which is closed with a rubber bung. A 1/4-in. copper pipe (C) is screwed through the bottom of the boiler, turns at right angles and ends in a machined brass jet (D) of internal diameter 1/16-in. This pipe is lagged outside the boiler with asbestos string. Heat is provided by burning 'Meta'* solid fuel in a small brass tray (E). A lagged handle (F) (more asbestos string) and a foot-flange (G) complete the supporting hardware.

The position of the pipe C in the boiling water, and the high conductivity of the material of which it is made, ensure, within the lagging, that the steam arrives at the jet at boiling point, and the emerging steam is colourless for a centimetre or two in air.

* 'Meta' is the registered trademark of Meta S. A. Basle.

In operation, a quarter-filling of water will give a long supply of steam, and will boil in a few minutes. Once boiling the fire is conveniently kept at the side of the tray nearest the handle by adding pieces of fuel there as needed, since in practice one mostly wants the jet directed slightly upwards. The pieces of plant that one does not want steamed may be protected from the steam by a piece of aluminium foil.

I gratefully acknowledge the help of Mr M. Kajtar who turned the idea into a reality, and has provided the drawing.

References

Aikman, D. P. and Anderson, W. P., 1971. A quantitative investigation of a peristaltic model for phloem translocation. *Ann. Bot. N.S.*, **35**, 761–72.

Bachofen, R. and Wanner, H., 1962. Transport und Verteilung von markierten Assimilate. II. Über die Transportbahnen von Assimilaten in Fruchtstielen von *Phaseolus*. *Planta*, **58**, 225–36.

Barrier, G. E. and Loomis, W. E., 1957. Absorption and translocation of 2,4-dichlorophenoxyacetic acid and P32 by leaves. *Pl. Physiol., Lancaster*, **32**, 225–31.

Bauer, L., 1949. Über den Wanderungsweg fluoreszierender Farbstoffe in den Siebröhren. *Planta*, **37**, 221–43.

  1953. Zur Frage der Stoffbewegungen in der Pflanze mit besonderer Berücksichtigung der Wanderung von Fluorochromen. *Planta*, **42**, 367–51.

Behnke, H.-D., 1968. Zum Feinbau der Siebelemente im Knoten der Dioscoreaceen. *Vortr. GesGeb. Bot.*, **2**, 20–3.

Behnke, H.-D. and Dörr, Inge, 1967. Zur Herkunft und Struktur der Plasmafilamente in Assimilatleitbahnen. *Planta*, **74**, 18–44.

Bennett, C. W., 1956. Biological relations of plant viruses. *A. Rev. Pl. Physiol.*, **7**, 143–70.

Biddulph, O. and Cory, R., 1957. An analysis of translocation in the phloem of the bean plant using THO, P^{32}, and C^{14}. *Pl. Physiol., Lancaster*, **32**, 608–19.

  1960. Demonstration of two translocation mechanisms in studies of bidirectional movement. *Pl. Physiol., Lancaster*, **35**, 689–95.

  1965. Translocation of C^{14} metabolites in the phloem of the bean plant. *Pl. Physiol., Lancaster*, **40**, 119–29.

Biddulph, O. and Markle, J., 1944. Translocation of radiophosphorus in the phloem of the cotton plant. *Am. J. Bot.*, **31**, 65–70.

Biddulph, Susann F., 1956. Visual indications of S^{35} and P^{32} translocation in the phloem. *Am. J. Bot.*, **43**, 143–8.

Bieleski, R. L., 1969. Phosphorus compounds in translocating phloem. *Pl. Physiol., Lancaster*, **44**, 497–502.

Birch-Hirschfeld, Luise, 1920. Untersuchungen über die Ausbreitungsgeschwindigkeit gelöster Stoffe in der Pflanze. *Jb. wiss. Bot.*, **59**, 171–262.

Böhning, R. H., Kendall, W. A. and Linck, A. J., 1953. Effect of temperature on growth and translocation in tomatoes. *Am. J. Bot.*, **40**, 150–3.

Böhning, R. H., Swanson, C. A. and Linck, A. J., 1952. The effect of hypocotyl temperatures on translocation of carbohydrates from bean leaves. *Pl. Physiol., Lancester*, **27**, 417–21.

Bonnemain, M. J.-L., 1965. Sur le transport diurne des produits d'assimilation lors de la floraison chez la tomate. *C.r. hebd. Séanc. Acad. Sci., Paris*, **260**, 2054–7.

Bowling, D. J. F., 1968. Translocation at 0 °C in *Helianthus annuus*. *J. exp. Bot.*, **19**, 381–8.

Bowmer, R. G., 1960. The influence of various factors on the rate of translocation of fluorescein in *Gossypium hirsutum* L. Dissertation for Ph.D. in the University of North Carolina.

Brown, H. T. and Escombe, F., 1905. Researches on some of the physiological processes of green leaves, with special reference to the interchange of energy between the leaf and its surroundings. *Proc. R. Soc., B* **76**, 29–111.

Brown, H. T. and Morris, G. H., 1893. A contribution to the chemistry and physiology of foliage leaves. *J. chem. Soc.*, **63**, 604–77.

Brown, K. J., 1968. Translocation of carbohydrate in cotton: movement to the fruiting bodies. *Ann. Bot.*, N.S. **32**, 703–13.

Büsgen, M., 1891. Der Honigtau. *Jena. Z. Naturw.*, **25**, 339–428.

Buvat, R., 1963*a*. Les infrastructures et la différenciation des cellules criblées de *Cucurbita pepo* L. *Port. Acta biol.*, A, **7**, 249–99.

1963*b*. Sur la présence d'acide ribonucléique dans les 'corpuscules muqueux' des cellules criblées de *Cucurbita pepo*. *C. r. hebd. Séanc. Acad. Sci., Paris*, **257**, 733–55.

Caldwell, J., 1930. Studies in translocation. I. Movement of food materials in the Swedish turnip. *Proc. R. Soc. Edinb.*, **50**, 130–41.

Canny, M. J., 1960*a*. The breakdown of sucrose during translocation. *Ann. Bot.*, N.S., **24**, 330–44.

1960*b*. The rate of translocation. *Biol. Rev.*, **35**, 507–32.

1961. Measurements of the velocity of translocation. *Ann. Bot.*, N.S., **25**, 152–67.

1962*a*. The translocation profile; sucrose and CO_2. *Ann. Bot.*, N.S., **26**, 181–96.

1962*b*. The mechanism of translocation. *Ann. Bot.*, N.S., **26**, 603–17.

Canny, M. J. and Askham, M. J., 1967. Physiological inferences from the evidence of translocated tracer: a caution. *Ann. Bot.*, N.S., **31**,

409–16.

Canny, M. J. and Markus, Katalin, 1960. Metabolism of phloem isolated from grapevine. *Aust. J. biol. Sci.*, **13**, 292–99.

Canny, M. J., Nairn, Barbara and Harvey, Margo, 1968. The velocity of translocation in trees. *Aust. J. Bot.*, **16**, 479–85.

Canny, M. J. and Phillips, O. M., 1963. Quantitative aspects of a theory of translocation. *Ann. Bot.*, N.S., **27**, 379–402.

Carr, D. J. and Wardlaw, I. F., 1965. The supply of photosynthetic assimilates to the grain from the flag leaf and ear of wheat. *Aust. J. biol. Sci.*, **18**, 711–19.

Cheadle, V. I. and Whitford, N. B., 1941. Observations on the phloem in the Monocotyledoneae. I. The occurrence and phytogenetic specialization in structure of the sieve tubes in the meta-phloem. *Am. J. Bot.*, **28**, 623–7.

Chen, S. L., 1951. Simultaneous movement of P^{32} and C^{14} in opposite directions in phloem tissue. *Am. J. Bot.*, **38**, 203–11.

Child, C. M. and Bellamy, A. W., 1919. Physiological isolation by low temperature in *Bryophyllum* and other plants. *Science, N.Y.*, **50**, 362–5.

Choi, I. C. and Aronoff, S., 1966. Photosynthate transport using tritiated water. *Pl. Physiol., Lancaster*, **41**, 1119–29.

Clauss, H., Mortimer, D. C. and Gorham, P. R., 1964. Time course study of translocation products of photosynthesis in Soybean plants. *Pl. Physiol., Lancaster*, **39**, 269–73.

Clements, H. F., 1934. Translocation of solutes in plants. *N.W. Sci.*, **8**, 9–21.

    1940. Movement of organic solutes in the sausage tree, *Kigelia africana*. *Pl. Physiol., Lancaster.*, **15**, 689–700.

Colwell, R. N., 1942. The use of radioactive phosphorus in translocation studies. *Am. J. Bot.*, **29**, 798–807.

Cooil, B. J., 1941. Significance of phloem exudate of *Cucurbita pepo*, with reference to translocation of organic materials. *Pl. Physiol., Lancaster*, **16**, 61–84.

Coulson, C. L. and Peel, A. J., 1968. Respiration of ^{14}C-labelled assimilates in stems of willow. *Ann. Bot.*, N.S., **32**, 867–76.

Crafts, A. S., 1931*a*. Movement of organic materials in plants. *Pl. Physiol., Lancaster*, **6**, 1–41.

    1931*b*. A technic for demonstrating plasmodesma. *Stain Technol.*, **6**, 127–9.

    1932. Phloem anatomy, exudation, and transport of organic nutrients in cucurbits. *Pl. Physiol., Lancaster*, **7**, 183–225.

    1933. Sieve-tube structure and translocation in the potato. *Pl. Physiol.,*

Lancaster, **8**, 81–104.

1938. Translocation in plants. *Pl. Physiol., Lancaster*, **13**, 791–814.

1939*a*. The relation between structure and function of the phloem. *Am. J. Bot.*, **26**, 172–7.

1939*b*. The protoplasmic properties of sieve tubes. *Protoplasma*, **33**, 389–98.

1939*c*. Movement of viruses, auxins, and chemical indicators in plants. *Bot. Rev.*, **5**, 471–504.

1951. Movement of assimilates, viruses, growth regulators, and chemical indicators in plants. *Bot. Rev.*, **17**, 203–84.

Crafts, A. S. and Crisp, C. E., 1971. *Phloem Transport in Plants*. Freeman, San Francisco.

Crafts, A. S. and Lorenz, O. A., 1944. Fruit growth and food transport in cucurbits. *Pl. Physiol., Lancaster*, **19**, 131–8.

Cronshaw, J. and Esau, Katherine, 1968*a*. P protein in the phloem of *Cucurbita*. I. The development of P-protein bodies. *J. Cell Biol.*, **38**, 25–39.

1968*b*. P protein in the phloem of *Cucurbita*. II. The P protein of mature sieve elements. *J. Cell Biol.*, **38**, 292–303.

Currier, H. B., 1957. Callose substance in plant cells. *Am. J. Bot.*, **44**, 478–88.

Currier, H. B., Esau, K. and Cheadle, V. I., 1955. Plasmolytic studies of phloem. *Am. J. Bot.*, **42**, 68–81.

Curtis, O. F., 1920. The upward translocation of foods in plants. I. Tissues concerned in translocation. *Am. J. Bot.*, **7**, 101–24.

1923. The effect of ringing a stem on the upward transfer of nitrogen and ash constituents. *Am. J. Bot.*, **10**, 361–82.

1925. Studies on the tissues concerned in the transfer of solutes in plants. The effect on the upward transfer of solutes of cutting the xylem as compared with that of cutting the phloem. *Am. J. Bot.*, **39**, 573–85.

1929. Studies on solute translocation in plants. *Am. J. Bot.*, **16**, 154–68.

Curtis, O. F. and Herty, S. D., 1936. The effect of temperature on translocation from leaves. *Am. J. Bot.*, **23**, 528–32.

Czapek, F., 1897. Über die Leitungswege der organischen Baustoffe in Pflanzenkörper. *Sber. Acad. Wiss. Wien.*, **106**, 117–170.

Dainty, J., Croghan, P. C. and Fensom, D. S., 1963. Electro-osmosis, with some applications to plant physiology. *Can. J. Bot.*, **41**, 953–66.

Deleano, N. C., 1911. Über die Ableitung der Assimilate durch die intakten, die chloroformierten und die plasmolysierten Blattstiele der Laubblätter. *Jb. wiss. Bot.*, **49**, 129–86.

de Vries, H., 1885. Ueber die Bedeutung der Circulation und der Rotation des Protoplasma fur den Stofftransport in der Pflanze. *Bot. Ztg.*, **43**, 1–6 and 17–26.

Dixon, H. H., 1923. Transport of organic substances in plants. *Nature, Lond.*, **110**, 547–51.

Dixon, H. H. and Ball, N. G., 1922. Transport of organic substances in plants. *Nature, Lond.*, **109**, 236–7.

Duloy, Margaret D. and Mercer, F. V., 1961. Studies in translocation. I. The respiration of phloem. *Aust. J. biol. Sci.*, **14**, 391–401.

Duloy, Margaret D., Mercer, F. V. and Rathgeber, Nele, 1961. Studies on translocation. II. Submicroscopic anatomy of the phloem. *Aust. J. biol. Sci.*, **14**, 506–18.

Engleman, E. M., 1965*a*. Sieve element of *Impatiens sultanii*. I. Wound reaction. *Ann. Bot.*, N.S., **29**, 83–101.

1965*b*. Sieve element of *Impatiens sultanii*. II. Developmental aspects. *Ann. Bot.*, N.S., **29**, 103–18.

Ervin, E. L. and Evert, R. F., 1967. Aspects of sieve element ontogeny and structure in *Smilax rotundifolia*. *Bot. Gaz.*, **128**, 138–44.

Esau, Katherine, 1950. Development and structure of phloem tissue. *Bot. Rev.*, **16**, 67–114.

1961. *Plants, Viruses and Insects*. Harvard University Press.

1965. Anatomy and cytology of *Vitis* phloem. *Hilgardia*, **37**, 17–72.

Esau, Katherine and Cheadle, V. I., 1958. Wall thickening in sieve elements. *Proc. natn. Acad. Sci. U.S.A.*, **44**, 546–53.

1959. Size of pores and their contents in sieve elements of dicotyledons. *Proc. natn. Acad. Sci. U.S.A.*, **45**, 156–62.

1961. An evaluation of studies on ultrastructure of sieve plates. *Proc. natn. Acad. Sci. U.S.A.*, **47**, 1716–26.

Esau, Katherine, Cheadle, V. I. and Risley, E. B., 1962. Development of sieve plate pores. *Bot. Gaz.*, **123**, 233–43.

Esau, Katherine and Cronshaw, J., 1968. Endoplasmic reticulum in the sieve element of *Cucurbita*. *J. Ultrast. Res.*, **23**, 1–14.

Esau, Katherine, Cronshaw, J. and Hoefert, L. L., 1967. Relation of beet yellows virus to the phloem and to movement in the sieve tubes. *J. Cell Biol.*, **32**, 71–87.

Esau, Katherine, Engleman, E. M. and Bisalputra, T., 1963. What are transcellular strands? *Planta*, **59**, 617–23.

Eschrich, W., 1953. Beiträge zur Kenntnis der Wundsiebröhrenentwicklung bei *Impatiens holsti*. *Planta*, **43**, 37–74.

1956. Kallose. (Ein kritischer Sammelbericht). *Protoplasma*, **47**, 487–530.

1961. Untersuchungen über den Ab- und Aufbau der Callose. *Z. Bot.*, **49**, 153–218.

1963. Beziehungen zwischen dem Auftreten von Callose und der Fein-struktur des primären Phloems bei *Cucurbita ficifolia*. *Planta*, **59**, 243–61.

1966. Translokation ^{14}C-markierter Assimilate im Licht und Dunkeln bei *Vicia faba*. *Planta*, **70**, 99–124.

1967. Bidirektionelle translokation in Siebröhren. *Planta*, **73**, 37–49.

1968. Translokation radioaktiv markierter Indolyl-3-essigsäure in Sie-bröhren von *Vicia faba*. *Planta*, **78**, 144–57.

Eschrich, W., Currier, H. B., Yamaguchi, S. and McNairn, R. B., 1965. Der einfluss verstärkter Callosebildung auf den Stofftransport in Siebröhren. *Planta*, **65**, 49–64.

Evans, L. T., Dunstone, R. L., Rawson, H. M. and Williams, R. F., 1970. The phloem of the wheat stem in relation to requirements for assimi-late by the ear. *Aust. J. biol. Sci.*, **23**, 743–52.

Evans, L. T. and Wardlaw, I. F., 1966. Independent translocation of ^{14}C-labelled assimilates and of the floral stimulus in *Lolium temulentum*. *Planta*, **68**, 310–26.

Evans, N. T. S., Ebert, M. and Moorby, J., 1963. A model for the trans-location of photosynthate in the Soybean. *J. exp. Bot.*, **14**, 221–31.

Evert, R. F. and Alfieri, F. J., 1965. Ontogeny and structure of coniferous sieve cells. *Am. J. Bot.*, **52**, 1058–66.

Evert, R. F., Eschrich, W., Medler, J. T. and Alfieri, F. J., 1968. Observa-tions on penetration of linden branches by stylets of the aphid *Longi-stigma caryae*. *Am. J. Bot.*, **55**, 860–74.

Evert, R. F. and Murmanis, Lidija, 1965. Ultrastructure of the secondary phloem of *Tilia americana*. *Am. J. Bot.*, **52**, 95–106.

Evert, R. F., Murmanis, Lidija and Sachs, I. B., 1966. Another view of the ultrastructure of *Cucurbita* phloem. *Ann. Bot.*, N.S., **30**, 563–85.

Fensom, D. S., 1957. The bioelectric potentials of plants and their func-tional significance. I. An electrokinetic theory of transport. *Can. J. Bot.*, **35**, 573–82.

Fensom, D. S., Clattenburg, R., Chung, T., Lee, D. R. and Arnold, D. C., 1968. Moving particles in intact sieve tubes of *Heracleum mantegaz-zianum*. *Nature, Lond.*, **219**, 531–2.

Fife, J. M., Price, C. and Fife, D. C., 1962. Some properties of phloem exudate collected from root of sugar beet. *Pl. Physiol., Lancaster*, **37**, 791–2.

Fischer, A., 1885. Studien über die Siebröhren der Dicotylenblätter. *Ber. Verh. sächs. Acad. Wiss. Math.-Phys. Kl.*, **36**, 245–90.

1886. Neue Beiträge zur Kenntnis der Siebröhren. *Ber. Verh. sächs. Acad. Wiss. Math.-Phys. Kl.*, **38**, 291–336.

Fischer, H., 1936–7. Untersuchungen über die Stickstoffwanderung in der höhrer Pflanze. *Z. Bot.*, **30**, 449–88.

Fisher, D. B., 1970. Kinetics of C-14 translocation in soybean. I. Kinetics in the stem. *Pl. Physiol., Lancaster*, **45**, 107–13.

Gage, R. S. and Aronoff, S., 1960. Translocation. III. Experiments with carbon 14, chlorine 36 and hydrogen 3. *Pl. Physiol., Lancaster*, **35**, 53–64.

Geiger, D. R., 1969*a*. Chilling and translocation inhibition. *Ohio J. Sci.*, **69**, 356–66.

    1969*b*. Effect of sink region cooling on translocation of photosynthate. *Pl. Physiol., Lancaster*, **41**, 1667–72.

Geiger, D. R., Saunders, M. A. and Cataldo, D. A., 1969. Translocation and accumulation of translocate in the sugar beet petiole. *Pl. Physiol., Lancaster*, **44**, 1657–65.

Geiger, D. R. and Sovonick, S. A., 1970. Temporary inhibition of translocation velocity and mass transfer rate by petiole cooling. *Pl. Physiol., Lancaster*, **46**, 847–9.

Geiger, D. R. and Swanson, C. A., 1965*a*. Sucrose translocation in the sugar beet. *Pl. Physiol., Lancaster*, **40**, 685–90.

    1965*b*. Evaluation of selected parameters in a sugar beet translocation system. *Pl. Physiol., Lancaster*, **40**, 942–7.

Goodall, D. W., 1946. The distribution of weight changes in the young tomato plant. II. Changes in dry weight of separated organs, and translocation rates. *Ann. Bot.*, N.S., **10**, 305–38.

Greenway, H. and Pitman, M. G., 1965. Potassium retranslocation in seedlings of *Hordeum vulgare*. *Aust. J. biol. Sci.*, **18**, 236–47.

Grew, N., 1682. *The Anatomy of Plants*. London.

Gunning, B. E. S. and Pate, J. S., 1969*a*. 'Transfer Cells'. Plant wall ingrowths, specialised in relation to short distance transport of solutes – their occurrence, structure and development. *Protoplasma*, **68**, 107–133.

    1969*b*. Cells with wall ingrowths (transfer cells) in the placenta of ferns. *Planta*, **87**, 271–4.

Gunning, B. E. S., Pate, J. S. and Briarty, L. G., 1968. Specialized 'transfer cells' in minor veins of leaves and their possible significance in phloem translocation. *J. Cell Biol.*, **37**, C7–C12.

Hale, C. R. and Weaver, R. J., 1962. The effect of developmental stage on direction of translocation of photosynthate in *Vitis vinifera*. *Hilgardia*, **33**, 89–131.

Hamilton, Susan and Canny, M. J., 1960. The transport of carbohydrates in Australian bracken. *Aust. J. biol. Sci.*, **13**, 479–85.

Hanstein, J., 1860. Über die Leitung des Saftes durch die Rinde und Folgerungen daraus. *Jb. wiss. Bot.*, **2**, 392–467.

Harel, Shulamith and Reinhold, Leonora, 1966. The effect of 2,4-dinitro-phenol on translocation in the phloem. *Physiol. Plantar.*, **19**, 634–43.

Hartig, T., 1858. Über die Bewegung des Saftes in den Holzpflanzen. *Bot. Ztg.*, **16**, 329–35 and 337–42.

1860. Beiträge zur physiologischen Forstbotanik. *Allg. Forst-u. Jagdztg.*, **36**, 257–61.

Hartt, Constance E., Kortschak, H. P., Forbes, Ada J. and Burr, G. O., 1963. Translocation of C^{14} in sugarcane. *Pl. Physiol., Lancaster*, **38**, 305–18.

Heine, H., 1885. Über die physiologische Funktion der Stärkescheide. *Ber. dt. bot. Ges.*, **3**, 189–94.

Hennig, E., 1966. Zur Histologie und Funktion von Einstichen der schwarzen Bohnenlaus (*Aphis fabae* Scop.) in *Vicia faba*-Pflanzen. *J. Insect. Physiol.*, **12**, 65–76.

Hewitt, S. P. and Curtis, O. F., 1948. The effect of temperature on loss of dry matter and carbohydrate from leaves by respiration and trans-location. *Am. J. Bot.*, **35**, 746–55.

Hill, A. W., 1901. The histology of the sieve-tubes of *Pinus*. *Ann. Bot.*, **15**, 575–611.

Hoad, G. V. and Peel, A. J., 1965. Studies on the movement of solutes between the sieve tubes and surrounding tissues in willow. I. Inter-ference between solutes and rate of translocation measurements. *J. exp. Bot.*, **16**, 433–51.

Horwitz, L., 1958. Some simplified mathematical treatments of transloca-tion in plants. *Pl. Physiol., Lancaster*, **33**, 81–93.

Hüber, B., 1932. Beobachtung und Messung pflanzlicher Saftströme. *Ber. dt. bot. Ges.*, **50**, 89–109.

1953. Die Gewinnung des Eschenmanna – eine Nutzung von Siebröhren-saft. *Ber. dt. bot. Ges.*, **66**, 340–6.

1957. Anatomical and physiological investigations on food transloca-tion in trees. In *The Physiology of Forest Trees*, ed. Thimann, K. V., pp. 367–79. Ronald, New York.

Hüber, B. and Rouschal, E., 1938. Anatomische und zellphysiologische Beobachtungen am Siebröhrensystem der Bäume. *Ber. dt. bot. Ges.*, **56**, 380–91.

Hüber, B., Schmidt, E. and Jahnel, H., 1937. Untersuchungen über die Assimilatstrom. I. *Tharandt. forstl. Jb.*, **88**, 1017–50.

Incoll, L. D. and Neales, T. F., 1970. The stem as a temporary sink before tuberization in *Helianthus tuberosus* L. *J. exp. Bot.*, **21**, 469–76.

Johnson, R. P. C., 1968. Microfilaments in pores between frozen-etched sieve elements. *Planta*, **81**, 314–32.

Jones, C. H., Edson, A. W. and Morse, W. J., 1903. The maple sap flow.

Bull. Vt. agric. Exp. Stn, **103**, 43–184.

Jones, H., Martin, R. V. and Porter, H. K., 1959. Translocation of [14]carbon in tobacco following assimilation of [14]carbon dioxide by a single leaf. *Ann. Bot.*, N.S., **23**, 493–508.

Joy, K. W., 1962. Transport of organic nitrogen through the phloem in sugar beet. *Nature, Lond.*, **195**, 618–19.

Kendall, W. A., 1952. The effect of intermittently varied petiole temperature on carbohydrate translocation from bean leaves. *Pl. Physiol., Lancaster*, **27**, 631–3.

　1955. Effect of certain metabolic inhibitors on translocation of P^{32} in bean plants. *Pl. Physiol., Lancaster*, **30**, 347–50.

Kennedy, J. S. and Mittler, T. E., 1953. A method of obtaining phloem sap via the mouth-parts of aphids. *Nature, Lond.*, **171**, 528.

Khan, A. A. and Sagar, G. R., 1966. Distribution of ^{14}C-labelled products of photosynthesis, during the commercial life of the tomato crop. *Ann. Bot.*, N.S., **30**, 727–43.

　1967. Translocation in tomato: the distribution of the products of photosynthesis of the leaves of a tomato plant during the phase of fruit production. *Hort. Res.*, **7**, 61–9.

　1969*a*. Alteration of the pattern of distribution of photosynthetic products in the tomato by manipulation of the plant. *Ann. Bot.*, N.S., **33**, 753–62.

　1969*b*. Changing patterns of distribution of the products of photosynthesis in the tomato plant with respect to time and to the age of a leaf. *Ann. Bot.*, N.S., **33**, 763–9.

King, R. W., Evans, L. T. and Wardlaw, I. F., 1968. Translocation of the floral stimulus in *Pharbitis nil* in relation to that of assimilates. *Z. Pflanzenphysiol.*, **59**, 377–88.

Kollmann, R., 1960. Untersuchungen über das Protoplasma der Siebröhren von *Passiflora coerulea*. II. Elektronoptische Untersuchungen. *Planta*, **55**, 67–107.

　1965. Zur Lokalisierung der Funktionstüchtigen Siebzellen im Sekundären Phloem von *Metasequoia glyptostroboides. Planta*, **65**, 173–9.

　1967. Autoradiographischer Nachweis der Assimilat-Transportbahn im sekundären Phloem von *Metasequoia glyptostroboides. Z. Pflanzenphysiol.*, **56**, 401–9.

　1968. Funktionelle Morphologie des Coniferen-Phloems. *Vortr. Ges-Geb. Bot.*, **2**, 15–19.

Kollmann, R. and Dörr, Inge, 1966. Lokalisierung funktionstüchtiger Siebzellen bei *Juniperus communis* mit Hilfe von Aphiden. *Z. Pflanzenphysiol.*, **55**, 131–41.

Kollmann, R. and Schumacher, W., 1963. Über die Feinstruktur des

Phloems von *Metasequoia glyptostroboides* und seine jahreszeitlichen Veränderungen. IV. Mitteilung. Weitere Beobachtungen zum Feinbau der Plasmabrücken in den Siebzellen. *Planta*, **60**, 360–89.

Kriedemann, P. E., 1970. The distribution of ^{14}C-labelled assimilates in mature lemon trees. *Aust. J. agric. Res.*, **21**, 623–32.

Kursanov, A. L., 1957. The root system of plants as the organ of compound exchange. *Izv. Akad. Nauk SSSR*, No. 6, 689–705.

1961. The transport of organic substances in plants. *Endeavour*, **20**, 19–25.

Kursanov, A. L. and Brovcenko, M. I., 1961. The influence of ATP on the entry of assimilates into the translocation system of sugar beet. *Fiziologiya Rast.*, **8**, 270–8.

Kursanov, A. L. and Turkina, M. V., 1952a. The respiration of vascular bundles. *D.A.N. SSSR*, **84**, 1073–6.

1952b. The respiration of translocating tissues and the transport of sucrose. *D.A.N. SSSR*, **85**, 649–52.

Lawton, J. R. S., 1967. Translocation in the phloem of *Dioscorea* spp. II. Distribution of translocates in the stem. *Z. Pflanzenphysiol.*, **58**, 8–16.

Lawton, J. R. S. and Canny, M. J., 1970. The proportion of sieve elements in the phloem of some tropical trees. *Planta*, **95**, 351–4.

Lecomte, H., 1889. Contribution à l'étude du liber des angiospermes. *Annls. Sci. nat. ser.*, **10**, 193–324.

Lee, K.-W., Whittle, Catherine M. and Dyer, H. J., 1966. Boron deficiency and translocation profiles in sunflower. *Physiologia Pl.*, **19**, 919–24.

Leonard, O. A., 1939. Translocation of carbohydrates in the sugar beet. *Pl. Physiol., Lancaster*, **14**, 55–74.

Lou, C.-H., Wu, S.-H., Chang, W.-C. and Shao, L.-M., 1957. Intercellular movements of protoplasm as a means of translocation of organic material in garlic. *Scientia sin.*, **6**, 139–57.

McNairn, R. B. and Currier, H. B., 1968. Translocation blockage by sieve plate callose. *Planta*, **82**, 369–80.

Mangham, S., 1917. On the mechanism of translocation in plant tissues. An hypothesis, with special reference to sugar conduction in sieve tubes. *Ann. Bot.*, **31**, 293–311.

1922. Transport of organic substances in plants. *Nature, Lond.*, **109**, 476–7.

Maskell, E. J. and Mason, T. G., 1929a. Studies on the transport of nitrogenous substances in the cotton plant. I. Preliminary observations on the downward transport of nitrogen in the stem. *Ann. Bot.*, **43**, 205–31.

1929b. Studies on the transport of nitrogenous substances in the cotton plant. II. Observations on concentration gradients. *Ann. Bot.*, **43**,

615–52.

1930*a*. Studies on the transport of nitrogenous substances in the cotton plant. III. The relation between longitudinal movement and concentration gradients in the bark. *Ann. Bot.*, **44**, 1–29.

1930*b*. Studies on the transport of nitrogenous substances in the cotton plant. IV. The interpretation of the effects of ringing, with special reference to the lability of the nitrogen compounds of the bark. *Ann. Bot.*, **44**, 233–67.

1930*c*. Studies on the transport of nitrogenous substances in the cotton plant. V. Movement to the boll. *Ann. Bot.*, **44**, 657–88.

Mason, T. G., 1926. Preliminary note on the physiological aspects of certain undescribed structures in the phloem of the greater yam, *Dioscorea alata. Linn. Scient. Proc. R. Dubl. Soc.*, **18**, 195–8.

Mason, T. G. and Lewin, C. T., 1926. On the rate of carbohydrate transport in the greater yam, *Dioscorea alata. Proc. R. Ir. Acad.*, **18**, 203–5.

Mason, T. G. and Maskell, E. J., 1928*a*. Studies on the transport of carbohydrates in the cotton plant. I. A study of diurnal variation in the carbohydrates of leaf, bark and wood, and of the effects of ringing. *Ann. Bot.*, **42**, 189–253.

1928*b*. Studies on the transport of carbohydrates in the cotton plant. II. The factors determining the rate and the direction of movement of sugars. *Ann. Bot.*, **42**, 571–636.

1931. Further studies on transport in the cotton plant. I. Preliminary observations on the transport of phosphorus, potassium and calcium. *Ann. Bot.*, **45**, 125–73.

1934. Further studies on transport in the cotton plant. II. An ontogenetic study of concentrations and vertical gradients. *Ann. Bot.*, **48**, 119–41.

Mason, T. G., Maskell, E. J. and Phillis, E., 1936*a*. Further studies on transport in the cotton plant. III. Concerning the independence of solute movement in the phloem. *Ann. Bot.*, **50**, 23–58.

1936*b*. Further studies on transport in the cotton plant. IV. On the simultaneous movement of solutes in opposite directions through the phloem. *Ann. Bot.*, **50**, 161–74.

Mason, T. G. and Phillis, E., 1936. Further studies on transport in the cotton plant. V. Oxygen supply and the activation of diffusion. *Ann. Bot.*, **50**, 455–99.

1937. The migration of solutes. *Bot. Rev.*, **3**, 47–71.

Mittler, T. E., 1953. Amino acids in the phloem sap and their excretion by aphids. *Nature, Lond.*, **172**, 207.

1957*a*. Studies on the feeding and nutrition of *Tuberolachnus salignus* (Gmelin) (Homoptera, Aphididae). I. The uptake of phloem sap. *J.*

exp. Biol., **34**, 334–41.

1957*b*. Sieve-tube sap via aphid stylets. In *The Physiology of Forest Trees*, ed. Thimann, K. V., pp. 401–5. Ronald, New York.

Moorby, J., Ebert, M. and Evans, N. T. S., 1963. The translocation of ¹¹C-labelled photosynthate in the Soybean. *J. exp. Bot.*, **14**, 210–20.

Moose, C. A., 1938. Chemical and spectroscopic analysis of phloem exudate and parenchyma sap from several species of plants. *Pl. Physiol., Lancaster*, **13**, 365–80.

Mortimer, D. C., 1965. Translocation of the products of photosynthesis in sugar beet petioles. *Can. J. Bot.*, **43**, 269–80.

Münch, E., 1926. Dynamik der Saftströmung. *Ber. dt. bot. Ges.*, **44**, 68–71.

1930. *Die Stoffbewegung in der Pflanze.* Jena.

Nagai, R. and Rebhun, L. I., 1966. Cytoplasmic microfilaments in streaming *Nitella* cells. *J. Ultrast. Res.*, **14**, 571–89.

Nägeli, C., 1861. Über die Siebröhren von *Cucurbita. Sber. bayer. Akad. Wiss.*, **1**, 212–38.

Nelson, C. D., Clauss, H., Mortimer, D. C. and Gorham, P. R., 1961. Selective translocation of products of photosynthesis in soybean. *Pl. Physiol., Lancaster*, **36**, 581–8.

Nelson, C. D. and Gorham, P. R., 1957*a*. Uptake and translocation of C¹⁴-labelled sugars applied to primary leaves of soybean seedlings. *Can. J. Bot.*, **35**, 339–47.

1957*b*. Translocation of radioactive sugars in the stems of soybean seedlings. *Can. J. Bot.*, **35**, 703–13.

1959*a*. Translocation of C¹⁴-labelled amino acids and amides in the stems of young soybean plants. *Can. J. Bot.*, **37**, 431–8.

1959*b*. Physiological control of the distribution of translocated amino acids and amides in young soybean plants. *Can. J. Bot.*, **37**, 439–47.

Nelson, C. D. and Krotkov, G., 1962. Preparing whole-leaf radioautographs for studying translocation. *Pl. Physiol., Lancaster*, **37**, 27–30.

Nelson, C. D., Perkins, H. J. and Gorham, P. R., 1958. Note on a rapid translocation of photosynthetically assimilated C¹⁴ out of the primary leaf of the young soybean plant. *Can. J. Biochem. Physiol.*, **36**, 1277–9.

1959. Evidence for different kinds of concurrent translocation of photosynthetically assimilated C¹⁴ in the soybean. *Can. J. Bot.*, **37**, 1181–89.

Nomoto, N. and Saeki, T., 1969. Dry matter accumulation in sunflower and maize leaves as measured by an improved half-leaf method. *Bot. Mag., Tokyo*, **82**, 20–27.

Nordhagen, R., 1954. Ethnobotanical studies on barkbread and the employment of wych-elm under natural husbandry. *Danm. geol. Unders.*, **80**, 262–308.

O'Brien, T. P. and McCully, Margaret E., 1969. *Plant Structure and Development*. Macmillan, New York.

1970. Cytoplasmic fibres associated with streaming and saltatory-particle movement in *Heracleum mantegazzianum*. *Planta*, **94**, 91–4.

O'Brien, T. P. and Thimann, K. V., 1966. Intracellular fibers in oat coleoptile cells and their possible significance in cytoplasmic streaming. *Proc. natn. Acad. Sci. U.S.A.*, **56**, 888–94.

O'Brien, T. P. and Wardlaw, I. F., 1961. The direct assay of ^{14}C in dried plant material. *Aust. J. biol. Sci.*, **14**, 361–7.

Othlinghaus, D., Schmitz, K. and Willenbrink, J., 1968. Zum assimilat-transport in wachsende Früchte von *Phaseolus vulgaris*. *Planta*, **80**, 89–95.

Palmquist, E. M., 1938. The simultaneous movement of carbohydrates and fluorescein in opposite directions in the phloem. *Am. J. Bot.*, **25**, 97–105.

Parker, B. C., 1965. Translocation in the giant kelp *Macrocystis*. I. Rates, direction, quantity of C^{14}-labelled products and fluorescein. *J. Phycol.*, **1**, 41–6.

Parker, J., 1964*a*. Sieve tube strands in tree bark. *Nature, Lond.*, **202**, 926–7.

1964*b*. Transcellular strands and intercellular particle movement in sieve tubes of some common trees. *Naturwissenschaften*, **11**, 273–4.

1965*a*. Strand characteristics in sieve tubes of some common tree species. *Protoplasma*, **60**, 86–93.

1965*b*. Stains for strands in sieve tubes. *Stain Technol.*, **40**, 223–5.

Pate, J. S., 1962. Root-exudation studies on the exchange of C^{14}-labelled organic substances between the roots and shoot of the nodulated legume. *Pl. Soil*, **17**, 333–56.

Pate, J. S. and Gunning, B. E. S., 1969. Vascular transfer cells in angiosperm leaves. A taxonomic and morphological survey. *Protoplasma*, **68**, 135–56.

Pate, J. S., Gunning, B. E. S. and Briarty, L. G., 1969. Ultrastructure and functioning of the transport system of the leguminous nodule. *Planta*, **85**, 11–34.

Pate, J. S. and O'Brien, T. P., 1968. Microautoradiographic study of the incorporation of labelled amino acids into insoluble compounds of the shoot of a higher plant. *Planta*, **78**, 60–71.

Peel, A. J., 1970. Further evidence for the relative immobility of water in sieve tubes of willow. *Physiologia Pl.*, **23**, 667–72.

Peel, A. J., Field, R. J., Coulson, C. L. and Gardner, D. C. J., 1969. Movement of water and solutes in sieve tubes of willow in response to puncture by aphid stylets, evidence against a mass flow of solution.

Physiologia Pl., **22**, 768–75.

Peel, A. J. and Weatherley, P. E., 1959. Composition of sieve-tube sap. *Nature, Lond.*, **184**, 1955–6.

1962. Studies in sieve-tube exudation through aphid mouth-parts: the effects of light and girdling. *Ann. Bot.*, N.S., **26**, 633–46.

Perkins, H. J., Nelson, C. D. and Gorham, P. R., 1959. A tissue-autoradiographic study of the translocation of C^{14}-labelled sugars in the stems of young soybean plants. *Can. J. Bot.*, **37**, 871–7.

Pfeiffer, M., 1937. Die Verteilung der osmotische Werte im Baum im Hinblick auf die Münsche Druckstromtheorie. *Flora, Jena*, **132**, 1–17.

Phillis, E. and Mason, T. G., 1933. Studies on the transport of carbohydrates in the cotton plant. III. The polar distribution of sugar in the foliage leaf. *Ann. Bot.*, **47**, 585–634.

1936. Further studies on transport in the cotton plant. VI. Interchange between the tissues of the corolla. *Ann. Bot.*, **50**, 679–97.

Priestley, J. H. and Wormall, A., 1925. On the solutes exuded by root pressure from vines. *New Phytol.*, **24**, 24–38.

Quinlan, J. D. and Sagar, G. R., 1962. An autoradiographic study of the movement of ^{14}C-labelled assimilates in the developing wheat plant. *Weed Res.*, **2**, 264–73.

Raciborski, M., 1898a. Ein Inhaltskörper des Leptoms. *Ber. dt. bot. Ges.*, **16**, 52–63.

1898b. Weitere Mitteilungen über das Leptomin. *Ber. dt. Ges.*, **16**, 119–23.

Ringoet, A., Sauer, G. and Gielink, A. J., 1968. Phloem transport of calcium in oat leaves. *Planta*, **80**, 15–20.

Roeckl, Brunhild, 1949. Nachweis eines Konzentrationshubs zwischen Palisadenzellen und Siebröhren. *Planta*, **36**, 530–50.

Rogers, A. W., 1967. *Techniques of Autoradiography.* Elsevier, Amsterdam.

Rouschal, E., 1941. Untersuchungen über die Protoplasmatik und Funktion der Siebröhren. *Flora, Jena*, **35**, 135–200.

Sachs, J., 1863. Über die Stoffe, welche das Material zum Wachstum der Zellhäute liefern. *Jb. wiss. Bot.*, **3**, 183–258.

1884. Ein Beitrag zur Kenntniss der Ernährungsthätigkeit der Blätter. *Arb. bot. Inst. Würzburg.*, **3**, 1–33.

Schmidt, E. W., 1917. *Bau und Funktion der Siebröhre der Angiospermen.* Jena.

Schmitz, K. and Willenbrink, J., 1967. Histoautoradiographischer Nachweis ^{14}C-markierter Assimilate im Phloem. *Z. Pflanzenphysiol.*, **58**, 97–107.

Schroeder, J., 1869–70. Beitrag zur Kenntnis der Frühjahrsperiode des Ahorns (*Acer platanoides*). *Jb. wiss. Bot.*, **7**, 261–343.

Schumacher, A., 1948. Beitrag zur Kenntnis des Stofftransportes in dem Siebröhrensystem höhrer Pflanzen. *Planta*, **35**, 642–700.

Schumacher, W., 1930. Untersuchungen über die Lokalisation der Stoffwanderung in den Leitbündeln höhrer Pflanzen. *Jb. wiss. Bot.*, **73**, 770–823.

1933. Untersuchungen über die Wanderung des Fluoreszeins in den Siebröhren. *Jb. wiss. Bot.*, **77**, 685–732.

1937. Weitere Untersuchungen über die Wanderung von Farbstoffen in den Siebröhren. *Jb. wiss. Bot.*, **85**, 422–49.

1939. Über die Plasmolysierbarkeit der Siebröhren. *Jb. wiss. Bot.*, **88**, 545–53.

1950. Bewegung des Fluoreszeins in den Siebröhren. *Planta*, **37**, 626–34.

Shear, C. B. and Faust, M., 1970. Calcium transport in apple trees. *Pl. Physiol., Lancaster*, **45**, 670–4.

Shih, C. Y. and Currier, H. B., 1969. Fine structure of phloem cells in relation to translocation in the cotton seedling. *Am. J. Bot.*, **56**, 464–72.

Sisler, E. C., Dugger, W. M. and Gauch, H. G., 1956. The role of boron in the translocation of organic compounds in plants. *Pl. Physiol., Lancaster*, **31**, 11–17.

Small, J., 1939. Technique for the observation of protoplasmic streaming in sieve tubes. *New Phytol.*, **38**, 176–7.

Spanner, D. C., 1958. The translocation of sugar in sieve tubes. *J. exp. Bot.*, **9**, 332–42.

1962. A note on the velocity and the energy requirement of translocation. *Ann. Bot.*, N.S., **26**, 511–16.

Spanner, D. C. and Prebble, J. N., 1962. The movement of tracers along the petiole of *Nymphoides peltatum*. *J. exp. Bot.*, **13**, 294–306.

Steucek, G. L. and Koontz, H. V., 1970. Phloem mobility of magnesium. *Pl. Physiol., Lancaster*, **46**, 50–2.

Steward, F. C. and Priestley, J. H., 1932. Movement of organic materials in plants. A note on a recently suggested mechanism. *Pl. Physiol., Lancaster*, **7**, 165–71.

Stout, P. R. and Hoagland, D. R., 1939. Upward and lateral movement of salt in certain plants as indicated by radioactive isotopes of potassium, sodium and phosphorus absorbed by roots. *Am. J. Bot.*, **26**, 320–4.

Strasburger, E., 1891. Über den Bau und die Verrichtungen der Leitungsbahnen in den Pflanzen. *Histol. Beitr.*, **3**.

Swanson, C. A. and Böhning, R. H., 1951. The effect of petiole temperature on the translocation of carbohydrates from bean leaves. *Pl. Physiol., Lancaster*, **26**, 557–64.

Swanson, C. A. and El Shishiny, E. D. H., 1958. Translocation of sugars in the Concord Grape. *Pl. Physiol., Lancaster*, **33**, 33–7.

Swanson, C. A. and Geiger, D. R., 1967. Time course of low temperature inhibition of sucrose translocation in sugar beets. *Pl. Physiol., Lancaster*, **42**, 751–6.

Swanson, C. A. and Whitney, J. B., 1953. Studies on the translocation of foliar-applied P^{32} and other radioisotopes in bean plants. *Am. J. Bot.*, **40**, 816–23.

Tammes, P. M. L., 1933. Observations on the bleeding of palm trees. *Recl. Trav. bot. néerl.*, **30**, 514–36.

1952. On the rate of translocation of bleeding-sap in the fruitstalk of *Arenga. Proc. Sect. Sci. K. ned. Acad. Wet.*, **55**, 141–3.

1958. Micro- and macro-nutrients in sieve-tube sap of palms. *Acta bot. néerl.*, **7**, 233–4.

Taylor, G. I., 1953. Dispersion of soluble matter in solvent flowing slowly through a tube. *Proc. R. Soc.*, A, **219**, 186–203.

1954. The dispersion of matter in turbulent flow through a pipe. *Proc. R. Soc.*, A, **223**, 446–68.

Thaine, R., 1961. Transcellular strands and particle movement in mature sieve tubes. *Nature, Lond.*, **192**, 772–3.

1962. A translocation hypothesis based on the structure of plant cytoplasm. *J. exp. Bot.*, **13**, 152–60.

1964a. Cytoplasm exudate from cut phloem. *Nature, Lond.*, **203**, 544–5.

1964b. Protoplast structures in sieve tube elements. *New Phytol.*, **63**, 236–43.

1964c. The protoplasmic-streaming theory of phloem transport. *J. exp. Bot.*, **15**, 470–84.

1965. Surface associations between particles and the endoplasmic reticulum in protoplasmic streaming. *New Phytol.*, **64**, 118–30.

1969. Movement of sugars through plants by cytoplasmic pumping. *Nature, Lond.*, **222**, 873–5.

Thaine, R., Ovenden, Stella L. and Turner, J. S., 1959. Translocation of labelled assimilates in the soybean. *Aust. J. biol. Sci.*, **12**, 349–72.

Thaine, R., Probine, M. C. and Dyer, P. Y., 1967. The existence of transcellular strands in mature sieve elements. *J. exp. Bot.*, **18**, 110–27.

Thoday, D., 1910a. Experimental researches on vegetable assimilation. V. A critical examination of Sachs' method for using increase of dry weight as a measure of carbon dioxide assimilation in leaves. *Proc. R. Soc.*, B., **82**, 1–55.

1910b. Experimental researches on vegetable assimilation. VI. Some experiments on assimilation in the open air. *Proc. R. Soc.*, B, **82**, 421–50.

Thrower, Stella L., 1962. Translocation of labelled assimilates in the soybean. *Aust. J. biol. Sci.*, **15**, 629–49.

1965. Translocation of labelled assimilates in soybean. IV. Some effects of low temperature on translocation. *Aust. J. biol. Sci.*, **18**, 449–61.

Tingley, Mary A., 1944. Concentration gradients in plant exudates with reference to the mechanism of translocation. *Am. J. Bot.*, **31**, 30–8.

Tomlinson, P. B. and Zimmermann, M. H., 1967. The 'wood' of monocotyledons. *Bull. Int. Ass. Wood Anatomists*, 1967/2, 4–24.

Trip, P. and Gorham, P. R., 1967. Autoradiographic study of the pathway of translocation. *Can. J. Bot.*, **45**, 1567–73.

1968*a*. Bi-directional translocation of sugars in sieve tubes of squash plants. *Pl. Physiol., Lancaster*, **43**, 877–82.

1968*b*. Translocation of sugar and tritiated water in squash plants. *Pl. Physiol., Lancaster*, **43**, 1845–9.

Trip, P., Nelson, C. D. and Krotkov, G., 1965. Selective and preferential translocation of C^{14}-labelled sugars in white ash and lilac. *Pl. Physiol., Lancaster*, **40**, 740–7.

Tsao, T.-H. and Liu, C.-Y., 1957. On the problem of transport of organic material in vascular tissue. (Chinese, English summary.) *Acta bot. sin.*, **6**, 269–80.

Ullrich, W., 1961. Zur Sauerstoffabhängigkeit des Transportes in den Siebröhren. *Planta*, **57**, 402–29.

1962. Zur Wirkung von Adenosintriphosphat auf den Fluoresceintransport in den Siebröhren. *Planta*, **57**, 713–17.

van den Honert, T. H., 1932. On the mechanism of transport of organic materials in plants. *Proc. Sect. Sci. K. ned. Acad. Wet.*, **35**, 1104–1111.

Vernon, L. P. and Aronoff, S., 1952. Metabolism of Soybean leaves. IV. Translocation from Soybean leaves. *Archs Biochem. Biophys.*, **36**, 383–98.

Wanner, H., 1952. Phosphataseverteilung und Kohlenhydrat-transport in der Pflanze. *Planta*, **41**, 190–4.

1953*a*. Die Zusammensetzung des Siebröhren-saftes: Kohlenhydrate. *Ber. schweiz. bot. Ges.*, **63**, 162–8.

1953*b*. Enzyme der Glykolyse im Phloemsaft. *Ber. schweiz. bot. Ges.*, **63**, 201–12.

Wardlaw, I.F., 1965. The velocity and pattern of assimilate translocation in wheat plants during grain development. *Aust. J. biol. Sci.*, **18**, 269–81.

1967. The effect of water stress on translocation in relation to photosynthesis and growth. I. Effect during grain development in wheat. *Aust. J. biol. Sci.*, **20**, 25–39.

Wark, Margaret C., 1965. Fine structure of the phloem of *Pisum sativum*.

II. The companion cell and phloem parenchyma. *Aust. J. Bot.*, **13**, 185–93.

Weatherley, P. E. and Johnson, R. P. C., 1968. The form and function of the sieve tube: a problem in reconciliation. *Int. Rev. Cytol.*, **24**, 149–92.

Weatherley, P. E., Peel, A. J. and Hill, G. P., 1959. The physiology of the sieve tube. Preliminary experiments using aphid mouth parts. *J. exp. Bot.*, **10**, 1–16.

Webb, J. A., 1967. Translocation of sugars in *Cucurbita melopepo*. IV. Effects of temperature change. *Pl. Physiol., Lancaster*, **42**, 881–5.

Webb, J. A. and Gorham, P. R., 1964. Translocation of photosynthetically assimilated ^{14}C in straight-necked squash. *Pl. Physiol., Lancaster*, **39**, 663–72.

1965. The effect of node temperature on assimilation and translocation of C^{14} in the squash. *Can. J. Bot.*, **43** (9), 1009–20.

Webb, K. L. and Burley, J. W. A., 1964. Stachyose translocation in plants. *Pl. Physiol., Lancaster*, **39**, 973–7.

Weevers, T. and Westenberg, J., 1931. Versuche zur Prüfung der Münschen Theorie der Stoffbewegungen in der Pflanze. *Proc. Sect. Sci. K. ned. Acad. Wet.*, **34**, 1173–8.

Weintraub, R. L. and Brown, J. W., 1950. Translocation of exogenous growth regulators in the bean seedling. *Pl. Physiol., Lancaster*, **25**, 140–9.

Weiser, C. J. and Blaney, L. T., 1964. The question of boron and sugar translocation in plants. *Physiologia Pl.*, **17**, 589–99.

Went, F. W., 1944. Plant growth under controlled conditions. III. Correlation between various physiological processes and growth in the tomato plant. *Am. J. Bot.*, **31**, 597–618.

Went, F. W. and Hull, H. M., 1949. The effect of temperature upon translocation of carbohydrates in the tomato plant. *Pl. Physiol., Lancaster*, **24**, 505–26.

Whittle, Catherine M., 1964a. Translocation in *Pteridium*. *Ann. Bot.*, N.S., **28**, 331–8.

1964b. Translocation and temperature. *Ann. Bot.*, N.S., **28**, 339–44.

Willenbrink, J., 1957. Über die Hemmung des Stofftransports in den Siebröhren durch lokale Inaktivierung verschiedener Atmungsenzyme, *Planta*, **48**, 269–342.

1966. Zur lokale Hemmung des Assimilattransports durch Blausäure. *Z. Pflanzenphysiol.*, **55**, 119–30.

1968. Einige Beziehungen zwischen Stoffwechsel und Ferntransport in Phloem. *Vortr. GesGeb. Bot.*, **2**, 42–9.

Willenbrink, J. and Kollmann, R., 1966. Über den Assimilattransport in

Phloem von *Metasequoia*. *Z. Pflanzenphysiol.*, **55**, 42–53.

Zech, H., 1952. Untersuchungen über den Infektionsvorgang und die Wanderung des Tabakmosaikvirus im Pflanzenkörper. *Planta*, **40**, 461–514.

Ziegler, H., 1956. Untersuchungen über die Leitung und Sekretion der Assimilate. *Planta*, **47**, 447–500.

1958. Über die Atmung und den Stofftransport in den isolierten Leitbündeln der Blattstiele von *Heracleum mantegazzianum* Somm. et Lev. *Planta*, **51**, 186–200.

Ziegler, H. and Kluge, M., 1962. Die Nucleinsäuren und ihre Bausteine in Siebröhrensaft von *Robinia pseudoacacia* L. *Planta*, **58**, 144–53.

Ziegler, H. and Mittler, T. E., 1959. Über den Zuckergehalt der Siebröhren-bzw. Siebzellensäfte von *Heracleum mantegazzianum* und *Picea abies* (L). Karst. *Z. Naturf.*, **14b**, 278–81.

Ziegler, H. and Schnabel, Margarete, 1961. Über Harnstoffderivate im Siebröhrensaft. *Flora, Jena*, **150**, 306–17.

Ziegler, H. and Vieweg, G. H., 1961. Der experimentell Nachweis einer Massenströmung im Phloem von *Heracleum mantegazzianum* Somm. et Lev. *Planta*, **56**, 402–15.

Zimmermann, M. H., 1957a. Translocation of organic substances in trees. I. The nature of the sugars in the sieve tube exudate of trees. *Pl. Physiol., Lancaster*, **32**, 288–91.

1957b. Translocation of organic substances in trees. II. On the translocation mechanism in the phloem of white ash (*Fraxinus americana* L.). *Pl. Physiol., Lancaster*, **32**, 399–404.

1958a. Translocation of organic substances in trees. III. The removal of sugars from the sieve tubes in the white ash (*Fraxinus americana* L.). *Pl. Physiol., Lancaster*, **33**, 213–17.

1958b. Translocation of organic substances in the phloem of trees. In *The Physiology of Forest Trees*, ed. Thimann, K.V., pp. 381–400. Ronald, New York.

1960. Longitudinal and tangential movement within the sieve-tube system of White Ash (*Fraxinus americana* L.). *Beih. Zn Schweiz. Forstver.*, **30**, 289–300.

1961. Movement of organic substances in trees. *Science, N.Y.*, **133**, 73–9.

1962. Translocation of organic substances in trees. V. Experimental double interruption of phloem in White Ash (*Fraxinus americana* L.). *Pl. Physiol., Lancaster*, **37**, 527–30.

1969. Translocation velocity and specific mass transfer in the sieve tubes of *Fraxinus americana* L. *Planta*, **84**, 272–8.

Author index

Subject index